The Spatial Turn

Across the disciplines, the study of space has undergone a profound and sustained resurgence. Space, place, mapping, and geographical imaginations have become commonplace topics in a variety of analytical fields in part because globalization has accentuated the significance of location. While this transformation has led to a renaissance in human geography, it also has manifested itself in the humanities and other social sciences. The purpose of this book is not to announce that space is significant, which by now is well known, but to explore how space is analyzed by a variety of disciplines, to compare and contrast these approaches, identify commonalities, and understand how and why differences appear.

The volume includes works by 13 scholars from a variety of geographical regions and disciplines. All have published about how space is used, represented, and given meaning in their respective fields. The chapters combine up-to-date literature reviews concerning the role of space in each discipline and several offer original empirical analyses. The introduction surveys the development of the spatial turn across the fields under consideration. Some chapters are concerned with Geography; others explore the role of space in contemporary Anthropology, English and Hispanic studies, Sociology, Religion, Political Science, Film, and Cultural Studies.

Despite frequent reference to the spatial turn, this is the first volume to explicitly address how theory and practice concerning space are used in a variety of fields from diverse conceptual perspectives. This book will appeal to everyone conducting conceptual and theoretical research on space, not simply in Geography, but in related fields as well.

Barney Warf is Professor of Geography at the University of Kansas. His research and teaching interests lie within the broad domain of human geography, particularly economic and political issues.

Santa Arias is Associate Professor of Spanish at the University of Kansas. She specializes in the literatures of colonial Latin America and interdisciplinary approaches to the study of literature and culture.

Routledge studies in human geography

This series provides a forum for innovative, vibrant, and critical debate within Human Geography. Titles will reflect the wealth of research which is taking place in this diverse and ever-expanding field.

Contributions will be drawn from the main sub-disciplines and from innovative areas of work which have no particular sub-disciplinary allegiances.

Published:

1. **A Geography of Islands**
   Small island insularity
   Stephen A. Royle

2. **Citizenships, Contingency and the Countryside**
   Rights, culture, land and the environment
   Gavin Parker

3. **The Differentiated Countryside**
   Jonathan Murdoch, Philip Lowe, Neil Ward and Terry Marsden

4. **The Human Geography of East Central Europe**
   David Turnock

5. **Imagined Regional Communities**
   Integration and sovereignty in the global south
   James D Sidaway

6. **Mapping Modernities**
   Geographies of Central and Eastern Europe 1920–2000
   Alan Dingsdale

7. **Rural Poverty**
   Marginalisation and exclusion in Britain and the United States
   Paul Milbourne

8. **Poverty and the Third Way**
   Colin C. Williams and Jan Windebank

9. **Ageing and Place**
   Edited by *Gavin J. Andrews and David R. Phillips*

10. **Geographies of Commodity Chains**
    Edited by *Alex Hughes and Suzanne Reimer*

11. **Queering Tourism**
Paradoxical performances at gay pride parades
Lynda T. Johnston

12. **Cross-Continental Food Chains**
Edited by *Niels Fold and Bill Pritchard*

13. **Private Cities**
Edited by *Georg Glasze, Chris Webster and Klaus Frantz*

14. **Global Geographies of Post Socialist Transition**
Tassilo Herrschel

15. **Urban Development in Post-Reform China**
Fulong Wu, Jiang Xu and Anthony Gar-On Yeh

16. **Rural Governance**
International perspectives
Edited by *Lynda Cheshire, Vaughan Higgins and Geoffrey Lawrence*

17. **Global Perspectives on Rural Childhood and Youth**
Young rural lives
Edited by *Ruth Panelli, Samantha Punch, and Elsbeth Robson*

18. **World City Syndrome**
Neoliberalism and inequality in Cape Town
David A. McDonald

19. **Exploring Post Development**
Aram Ziai

20. **Family Farms**
Harold Brookfield and Helen Parsons

21. **China on the Move**
Migration, the State, and the Household
C. Cindy Fan

22. **Participatory Action Research Approaches and Methods**
Connecting People, Participation and Place
Sara Kindon, Rachel Pain and Mike Kesby

23. **Time-Space Compression**
Historical Geographies
Barney Warf

24. **Sensing Cities**
Monica Degen

25. **International Migration and Knowledge**
Allan Williams and Vladimir Baláž

26. **The Spatial Turn**
Interdisciplinary Perspectives
Edited by *Barney Warf and Santa Arias*

Not yet published:

27. **Design Economies and the Changing World Economy**
Innovation, Production and Competitiveness
John Bryson and Grete Rustin

28. **Whose Urban Renaissance?**
An International Comparison of Urban Regeneration Policies
Libby Porter and Katie Shaw

29. **Tourism Geography**
A New Synthesis Second Edition
Stephen Williams

30. **Critical Reflections on Regional Competitiveness**
Gillian Bristow

The Spatial Turn

Interdisciplinary perspectives

Edited by
Barney Warf
and
Santa Arias

LONDON AND NEW YORK

First published 2009
by Routledge
2 Park Square, Milton Park, Abingdon, Oxfordshire OX14 4RN

Simultaneously published in the USA and Canada
by Routledge
711 Third Avenue, New York, NY 10017
First issued in paperback 2014

Routledge is an imprint of the Taylor and Francis Group, an informa business

Typeset in Times New Roman by
Book Now Ltd, London

British Library Cataloguing in Publication Data
A catalogue record for this book is available from the British Library

Library of Congress Cataloging in Publication Data
The spatial turn: interdisciplinary perspectives/[edited by]
Barney Warf and Santa Arias.
p. cm.
Includes bibliographical references.
1. Geography—Philosophy. 2. Geography—Social aspects. 3. Human
geography. I. Warf, Barney, 1956—II. Arias, Santa.
G70.S685 2008
910.01—dc22 2008010331

ISBN 978-0-415-77573-1 (hbk)
ISBN 978-0-415-76221-2 (pbk)
ISBN 978-0-203-89130-8 (ebk)

Contents

Illustrations

Table

Figures

Contributors

Santa Arias, Spanish, University of Kansas. She is the author of *Retórica, historia y polémica: Bartolomé de las Casas y la tradición intelectual renacentista* (2001) and has co-edited *Mapping Colonial Spanish America* (2002) and *Approaches to Teaching the Writings of Bartolomé de las Casas* (2008). She is completing the book *Spaces of Conversion: Writing and Mapping the Spiritual Conquest in Colonial Latin America*. Her research is devoted to the role of space in the material and discursive understanding of Latin American colonial societies.

Sebastián Cobarrubias, Geography, University of North Carolina-Chapel Hill. He is a Ph.D. candidate in Geography. His dissertation is entitled *Navigating a Changing 'Europe' from Below: Activist Cartographies by Social Movements in 'Spain'*. His publications include chapters in *An Atlas of Radical Cartography* (2007) and *Constituent Imagination: Militant Investigation, Collective Theorization* (2006). Helped to co-found the prolific Counter Cartographies Collective at UNC-CH.

John Corrigan, Edwin Scott Gaustad Professor of Religion and Professor of History. He is the author or co-author of a dozen books on religion in America, the history of monotheism, religion and colonialism, and religion and emotion. He recently has published *The Oxford Handbook of Religion and Emotion* (2008) and co-authored with Lynn Neal, *Religious Intolerance in America: A Documentary History,* which is forthcoming from the University of North Carolina Press.

Harry F. Dahms, Sociology, University of Tennessee. He has edited the volume *Transformations of Capitalism: Economy, Society and the State in Modern Times* (2000) for the Main Trends in the Modern World series and has served as guest editor for the annual *Current Perspectives in Social Theory* since 2006. In addition, he has published many articles in distinguished journals in his field. At present, Dahms is working on the book manuscript, *Delivering Society: A Dynamic Theory of Modern Capitalism*.

Pamela K. Gilbert, English, University of Florida. She is author of *Disease, Desire and the Body in Victorian Women's Popular Novels* (1997) and *Mapping* *the Victorian Social Body* (2004). In addition, she published an edited collection entitled *Imagined Londons* (2002); co-edited *Beyond Sensation: Mary Elizabeth Braddon in Context* (1999), and is series editor of SUNY Press series, Studies in the Long Nineteenth Century. She is currently working on a book on the construction of the social body and cholera in England, 1832–66.

Jeffrey Kopstein, Political Science, University of Toronto. He has research interests in comparative politics, historical political economy, and European politics. He published *The Politics of Economic Decline in East Germany, 1945–1989* (1997), and co-edited *Comparative Politics: Identities, Institutions, and Interests in a Changing Global Order* (2000). Recent articles have appeared in *World Politics* (1996, 2000); *Political Theory* (2001); *German Politics and Society* (2002); *Comparative Politics* (2003); and *Slavic Review* (2003); *Theory and Society* (2005); and *Canadian Journal of Political Science* (2005).

Mariselle Meléndez, Latin American Literature, University of Illinois, Urbana-Champaign. Her publications in the field of Latin American colonial literature and culture include *Raza, género e hibridez en El lazarillo de ciegos caminantes* (1999), the edited volume *Mapping Colonial Spanish America* (2002), and two dozen articles in prestigious journals in Hispanic and Latin American cultural studies. Her research interests encompass the representation of space and bodies as they interconnect with race, ethnicity, and gender issues in colonial studies.

John Pickles, Earl N. Phillips Distinguished Chair of International Studies, University of North Carolina-Chapel Hill. He authored and edited many books and articles, including *State and Society in Post-Socialist Economies* (2007); A *History of Spaces: Cartographic Reason, Mapping, and the Geo-Coded World* (2004); *Environmental Transitions: Post-Communist Transformations and Ecological Defense in Central and Eastern Europe* (2000, with Petr Pavlinek); *Theorising Transition: The Political Economy of Post-Communist Transformations* (1998 with Adrian Smith)*; Ground Truth: The Social Implications of Geographical Information Systems* (1995); *Commonplaces, Humanism, and Geography* (1989); and *Phenomenology, Science, and Geography: Space and the Human Sciences* (1985). Helped to co-found the prolific Counter Cartographies Collective at UNC-CH.

Joan Ramon Resina, Spanish, Stanford University. He specializes in literature and cultural theory. His books include *La búsqueda del Grial* (1988), *Un sueño de piedra* (1990), *Los usos del clásico* (1991), *El cadáver en la cocina* (1997), *El postnacionalisme en el mapa global* (2005) and *Barcelona's Vocation of Modernity* (2008). He has edited six volumes and published one hundred articles and book chapters.

Margarita Serje, Anthropology, Universidad de los Andes (Colombia). She specializes in anthropological thought, development, globalization, and urban spaces and has published numerous books, including most recently *El Revés de la Nación: Territorios salvajes, fronteras y tierras de nadie* (2005) and *Palabras para desarmar: Una aproximación crítica al vocabulario del reconocimiento cultural en Colombia* (2002).

Edward W. Soja, Urban Planning, UCLA, distinguished scholar who has written extensively concerning poststructuralism in geography and the spatial turn. He wrote several highly influential volumes, including *Postmodern Geographies* (1989), *Thirdspace* (1996), and *Postmetropolis* (2000). In addition, he co-edited with A. J. Scott, *The City: Los Angeles and Urban Theory at the End of the Twentieth Century* (1996).

Barney Warf, Geography, University of Kansas. He has a long-standing interest in the spatiality of social theory and political economy, has published over ninety refereed articles in geography journals, and has co-authored and/or co-edited eight books, including most recently *The Encyclopedia of Human Geography* (2006) and *Time-Space Compression: Historical Geographies* (2008). Much of his work concerns services and information technologies.

Acknowledgements

Many people have helped in various ways to put this volume together. The editors would like to thank the ten contributors who produced such provocative essays, in particular Edward Soja for his inspirational scholarship that elevated space and place to the center of thought in the humanities and social sciences. We sincerely hope that this volume is useful to all of those working on issues of spatiality wherever they might be. The spatial turn is surely not confined to the disciplines represented here and we recognize that there are many unsung heroes who shed new light on space in their respective fields.

Our gratitude goes to Andrew Mould and the staff at Routledge who supported and guided this project from the beginning. We also would like to acknowledge the following institutions and individuals who have given permission to reproduce images and graphics: Lucy Patrick and the Robert Manning Strozier Library Special Collections at Florida State University; Rare Book and Manuscript Library at the University of Illinois at Urbana-Champaign; Patrimonio Nacional de España (Biblioteca del Real Palacio de Madrid); Fundación Pro-Sierra Nevada de Santa Marta; and Professor Tom Conley (Harvard University).

Finally, we deeply appreciate the support and encouragement of our families, friends, colleagues and graduate students. Our best wishes go to all of you.

1 Introduction: the reinsertion of space into the social sciences and humanities

Barney Warf and Santa Arias

> The geographical imagination is far too pervasive and important a fact of intellectual life to be left alone to geographers.
>
> (Harvey 1995: 161)

Human geography over the last two decades has undergone a profound conceptual and methodological renaissance that has transformed it into one of the most dynamic, innovative and influential of the social sciences. The discipline, which long suffered from a negative popular reputation as a trivial, purely empirical field with little analytical substance, has moved decisively from being an importer of ideas from other fields to an exporter, and geographers are increasingly being read by scholars in the humanities and other social sciences. As a result of the rebirth in scholarship in geography, other disciplines have increasingly come to regard space as an important dimension to their own areas of inquiry. Cosgrove (1999: 7), for example, argues that "A widely acknowledged 'spatial turn' across arts and sciences corresponds to post-structuralist agnosticism about both naturalistic and universal explanations and about single-voiced historical narratives, and to the concomitant recognition that position and context are centrally and inescapably implicated in all constructions of knowledge." Recent works in the fields of literary and cultural studies, sociology, political science, anthropology, history, and art history have become increasingly spatial in their orientation. From various perspectives, they assert that space is a social construction relevant to the understanding of the different histories of human subjects and to the production of cultural phenomena. In some ways, this transformation is expressed in simple semantic terms, i.e., the literal and metaphorical use and assumptions of "space," "place," and "mapping" to denote a geographic dimension as an essential aspect of the production of culture. In other ways, however, the spatial turn is much more substantive, involving a reworking of the very notion and significance of spatiality to offer a perspective in which space is every bit as important as time in the unfolding of human affairs, a view in which geography is not relegated to an afterthought of social relations, but is intimately involved in their construction. Geography matters, not for the simplistic and overly used reason that everything happens in space, but because *where* things happen is critical to knowing *how* and *why* they happen.

This volume charts this rise in spatial scholarship across several disciplines. We seek to explore how geographers have influenced other fields of scholarship and the many forms in which geography has motivated scholars to think spatially. With space and place at the center of the analytical agenda, geographical thought has arguably played a major role in helping to facilitate interdisciplinary inquiry that offers a richer, more contextualized understanding of human experience, social relations and the production of culture. Our goal in bringing together these authors and essays is to provide the reader with a sense of how space has entered into a variety of domains of knowledge. To be sure, as different disciplines have taken up geography in their own way they bring to bear their respective assumptions, languages, paradigms, applications, and examples about the meaning of the spatial. Thus, as the spatial turn has unfolded across the social sciences and humanities, the term has come to embrace an ever-larger set of uses and implications. But, conversely, space can serve as a window into different disciplines, a means of shedding light on what separates and what unites them. Because so many lines of thought converge on the topic of spatiality, space is a vehicle for examining what it means to be interdisciplinary or multidisciplinary, to cross the borders and divides that have organized the academic division of labor, to reveal the cultures that pervade different fields of knowledge, and to bring these contrasting lines of thought into a productive engagement with one another.

To appreciate the spatial turn, it is incumbent to know something about the changing role of space in the social sciences and humanities. The following section briefly charts this complex issue, noting the rise of historicism under modern capitalism and its decline under postmodernism. Next, the chapter asks why is it that space should be so important to understanding the contemporary world, and offers insights gleaned from analyses of globalization, cyberspace, changes in identity and subjectivity, and environmental issues. Finally, the chapter concludes by presenting a brief synopsis of each chapter.

The fall and rise of spatiality

In the nineteenth century, space became steadily subordinated to time in modern consciousness, a phenomenon that reflected the enormous time-space compression of the industrial revolution; intellectually, this phenomenon was manifested through the lens of historicism, a despatialized consciousness in which geography figured weakly or not at all, or, as Soja (1993: 140) defines it, as "an overdeveloped historical contextualization of social life and social theory that actively submerges and peripheralises the geographical or spatial imagination." Typically, historicist thought linearized time and marginalized space by positing the existence of temporal "stages" of development, a view that portrayed the past as the progressive, inexorable ascent from savagery to civilization, simplicity to complexity, primitiveness to civilization, and darkness to light. a trend made most explicit in Whiggish accounts of history. Likewise, historicists such as Hegel, Marx, and Toynbee offered sweeping teleological accounts that paid little attention to space, human consciousness, or the contingency of social life. In the same vein, Social Darwinism usurped the original theory of evolution as contingent and open-ended, substituting it with a simplistic, racist, linear view of phyletic gradualism such as Spencer's "survival of the fittest." Orientalist thought structured the Western geographical imagination such that distance from Europe became equated with increasingly more primitive stages of development, conflating continents with races in terms that were hierarchically organized in terms of their degree of alleged degree of temporal progress. In this way did historicism eclipse space in the service of imperial thought: beyond Europe was *before* Europe (McGrane 1983: 94), a theme articulated over and over again in modernization theory and its current neoliberal variants. All of these maneuvers robbed the understanding of social change of any sense of contingency, framing the past as a train of events leading inevitably to the present.

The reassertion of space into modern consciousness was a long, slow, and painful undertaking. In the 1920s, the Chicago School of sociologists and geographers attempted to inject space into urban analysis, a project that was poignant in its sensitivity to the experience of recent immigrants and the textures of ethnic neighborhoods and simultaneously doomed by its simplistic understanding of class, gender, power, and the world system. In the tumultuous 1960s, as Soja notes in his chapter, the seminal works of Henri Lefebvre (1974/1991) and Michel Foucault (1972/1980) were critical in suggesting that the organization of space was central to the structure and functioning of capitalism as a coherent whole. Moreover, Lefebvre maintained that space must be understood not simply as a concrete, material object, but also as an ideological, lived, and subjective one.

It was the injection of social theory – specifically Marxism, initially via the works of David Harvey (1973, 1982, 1985, 1989, 1990, 2006) – that formed the centerpiece for a critical re-evaluation of space and spatiality in social thought. Social theory repositioned the understanding of space from given to produced, calling attention to its role in the construction and transformation of social life and its deeply power-laden nature. Freed from the frozen geometries of spatial analysis, the plasticity of space rose to the fore, its contingent creation as a central moment in the reproduction of social life. Harvey's spatialization of Marxism, and concurrent Marxification of space, centered on the deep structure of commodity production and the conversion of commodities into money, generating a model of production and the labor process that shed light on its transformation of time and space. Landscapes, in this reading, reflected the logic of commodity production at any given historical moment, constituting a "spatial fix" or window of stability that enabled the process of commodity production to unfold unproblematically, at least for a fixed window of time. Eventually, however, capitalists as a whole are compelled to speed up the turnover time of capital, which is "the time of production together with the time of circulation of exchange" (Harvey 1989: 229). The resulting need to "annihilate space by time" – to substitute one spatial fix with another – is thus fundamental to the operation and survival of capitalism, i.e., its ability to reproduce itself at ever expanded spatial scales and to accelerate temporal rhythms of capital accumulation. In this way, capitalism exhibits a fundamental contradiction between fixity – the need to stabilize production temporarily in order to realize surplus value – and motion, the need to annihilate old geographies in an act of creative destruction and replace them with new, more efficient landscapes amenable to more recent systems of production. Capitalists must negotiate (often unsuccessfully) the knife-edge between using up old spaces and creating new ones.

This line of thought was picked up and advanced by Ed Soja (1989, 1996, 2000), whose works repeatedly and emphatically insisted that the spatial could not be subordinated to time or the social. Thus, he maintains that social theory should rest on the triangular foundations of time, space, and social structure, each of which contingently structures and is structured by the others. Thus, Soja (1993) argues that the spatial turn has involved the end of historicism, which privileged time over space, and the reassertion of space into social theory. In a sense, this trend marks a return to Kant's position that held the two dimensions to be of equal significance.

Similarly, Manuel Castells (1996, 1997), noting the pronounced degree to which postmodern capitalism relies on information as its primary resource, distinguished earlier *information* societies, in which productivity was derived from access to energy and the manipulation of materials, from later *informational* societies that emerged in the late twentieth century, in which productivity is derived primarily from knowledge and information. In his reading, the spatiality of postmodernism was manifested in the global "space of flows." He notes, for example, that while people live in places, postmodern power is manifested in the linkages among places, their interconnectedness, as personified by business executives shuttling among global cities and using the internet to weave complex geographies of knowledge invisible to almost all ordinary citizens. This process was largely driven by the needs of the transnational class of the powerful employed in information-intensive occupations.

Increasingly, in the 1980s, the reassertion of space came to embrace various aspects of human subjectivity, everyday life, and the multiple dimensions of identity that are central to any coherent understanding of social life. Giddens's (1984) theory of structuration, for example, transformed the once-strict dualism of structure and agency into a fluid duality, in which individuals, forming their biographies in time and space through the routines of everyday life, reproduce and transform their social worlds primarily without meaning to do so. Everyday thought and behavior, the unacknowledged preconditions to action, do not simply mirror the world; they constitute it as the outcomes to action. Social structures and relations are thus reproduced, and hence simultaneously changed, by the people who make them; individuals are both produced by, and producers of, history and geography. Given this logic, space could no longer be seen simply as a backdrop against which life unfolds sequentially, but rather, intimately tied to lived experience. As Foucault suggests, space "takes the form of relations among sites" (quoted in Soja 1996: 156).

Why space, why now?

The spatial turn is hardly the product of a few ivory tower intellectuals. Rather, this shift in social thought reflects much broader transformations in the economy, politics, and culture of the contemporary world. Such a view asserts that we cannot comprehend the production of spatial ideas independent of the production of spatiality, i.e., views of geography are only comprehensible by appeal to social and spatial context. Several forces have intersected since the late twentieth century to elevate space to new levels of material and ideological significance.

Contemporary globalization has undermined commonly held notions of Euclidean space by forming linkages among disparate producers and consumers intimately connected over vast distances through flows of capital and goods. Globalization is an annoyingly ambiguous term, however, and refers to a variety of processes that play out in different ways, from global cities to international trade to the internationalization of culture and consumption. Unprecedented volumes of immigration across national borders have increasingly confronted residents in many countries, particularly in Europe and North America, with issues of cultural difference. Tourism likewise allows for the selective appropriation of distant cultures and places. The media, increasingly under the control of a handful of robber barons, relay news and events across a global stage. The offshoring of many jobs from the developed world to the developing one has called attention to national differentials in skills and labor costs. International finance, a worldwide space of flows, global deregulation, and the decline in transport costs have all conspired to erode conventional geographic preoccupations with proximity. Far from annihilating the importance of space, globalization has increased it. Ironically, just as several pundits announced the "death of distance" and the "end of geography," geography acquired a renewed significance in the analysis of international flows of information, culture, capital, and people. As neoliberal capital operates ever more effortlessly on a worldwide stage, small differences among regions become increasingly important. Moreover, as numerous scholars have shown, globalization does not play out identically in different places (Swyngedouw 1997); rather, context matters, and the incorporation of unique places into the global division of labor changes not only those locales, but the world system as well. Glocalization, therefore, is a two-way street.

The rise of cyberspace and the internet has also raised issues of spatiality in several fields. The internet, an unregulated electronic network connecting an estimated one billion people in more than 180 countries in 2007, allows users to transcend distance virtually instantaneously. Telecommunications systems have become the central technology of postmodern capitalism, vital not only to corporations large and small, but also to consumption, personal communication, entertainment, education, politics, and numerous other domains of social life. Indeed, for many people who spend a great deal of time in the digital world, cyberspace has become such an important part of everyday life that the once-solid boundary between the real and the virtual has essentially dissolved: it is difficult today to tell where one ends and the other begins. In allowing people and firms to "jump scale," to connect effortlessly with others around the world at the click of a mouse, cyberspace has been instrumental to the production of complex, fragmented, jumbled spaces of postmodernity, all of which have called for mounting scrutiny. Similarly, Geographical Information Systems (GIS) offer dramatic new means to analyze spatiality, even if these are often drenched in empiricist and positivist epistemologies. GIS, therefore, is not simply reflective of the new importance of space, but also constitutive of it.

Globalization, the international media, and migration across national borders – all reflective of one underlying transformation – have also entailed a profound shift in identity and subjectivity (e.g. Featherstone 1990; Robertson 1992; Gubrium and Holstein 2000). Rather than a fixed, unified identity that lies at the core of the modern self, for example, there are grounds for arguing many people, hooked into different locales via the internet, consist of multiple, shifting, even contradictory "selves" who lie at the changing intersections of different language games. Jameson (1991: 14), for example, argues that "the alienation of the subject is displaced by the fragmentation of the subject." In this context, increasingly, the notion of the autonomous subject standing apart from the world he/she observes has come under question, and in its place lie a greater pluralistic affirmation of cultural difference based on numerous axes of identity (gender, ethnicity, sexuality, etc.).

Finally, the rapidly rising seriousness of global ecological and environmental problems has itself played no small role in elevating the significance of space. Issues that once could be understood and contained within relatively localized contexts, such as water pollution, are increasingly viewed as approachable only on a worldwide basis. Acid rain crosses national borders with ease. The enormous crisis of biodiversity unfolding across the world is inseparable from global patterns of habitat destruction, deforestation, and resource consumption. The gradual end of the age of cheap petrochemicals has added new urgency to this issue, and threatens to raise the "friction of distance" of commuting fields. And of course, looming over all of this is the dark shadow of global warming: as evidence mounts daily of rapidly melting icecaps and glaciers, of ecosystems turned upside down overnight, the location of sources of greenhouse gases has become a vitally problematic issue. Because environmental issues are unevenly distributed across space, and because geography as a discipline has a long history of investigating human–environment interactions, space and spatiality have become crucial dimensions in understanding and tackling these problems.

A sketch of this volume

The 11 essays that follow offer a variety of approaches to the revival of spatial scholarship. Some are analytical surveys of the literature in their respective disciplines, while others offer original empirical analyses of space in various contexts. Far from constituting some homogenous whole, they emphasize distinctly different interpretations of what spatiality means, how it is constructed, why it changes, and its implications for how society is organized and reorganized, knowledge constructed, and the possibilities and perils of change. Such lines of thought focus on a multitude of spatial scales, ranging from the most intimate, the body, to that of community, region, nation, and the entirety of the globe. Space as a swirling, complex, contingent, ever-changing maelstrom of possibilities is seen to be wrapped up in issues and contexts as diverse as Neolithic cities, eighteenth-century representations of the Americas by Europeans and Creoles, twenty-first-century Manhattan, the Cao Dai religion in Vietnam, post-Soviet Eurasia, cyberspace, the pre-Columbian Andes, and Spanish cinema. What gives these essays coherence, what unites them as a common point of departure, is the insistence that no social or cultural phenomenon can be torn from its spatial context, that geography is not some subordinate afterthought to history in the construction of social life, that no meaningful understanding of how human beings produce and reproduce their worlds can be achieved without invoking a sense that the social, the temporal, the intellectual, and the personal are inescapably always and everywhere also the spatial.

The collection begins with one of the key theorists in initiating the spatial turn, Edward Soja, who has long been one of the most aggressively spatial thinkers in the social sciences and influential across the disciplines. In an essay that weaves between his autobiographical journey and the broader intellectual history of contemporary spatiality, in which he charts his own, highly personal spatial turn by juxtaposing it with the transformation of social thought more broadly, he also argues that most of the spatial turn occurred outside of the discipline of Geography, the discipline for which space has long been the central, defining dimension. For Soja, spatiality has long been synonymous with the urban – in his case, a voyage that included the Bronx, Chicago, and Los Angeles – much as spatiality has exerted an effect that extended from Neolithic Catalhuyuk to the contemporary global city.

The depiction and inscription of space have increasingly come to form an important element in popular struggles against neoliberalism and reactionary globalization. Sebastián Cobarrubias and John Pickles chart the rise of subversive cartographies, cognitive maps that portray the profoundly disorienting reterritorializations of the post-Fordist age, and in the process simultaneously reflect the newfound role of space in geographies of resistance and help to call it into being. They focus on the efforts of two groups, the Bureau d'Études in Paris and Hackitectura in Seville, to construct emancipatory mapping strategies in the face of an increasingly neoliberal Europe. In the process of drawing together diverse political strands of civil society, they re-present space in forms that undermine the taken-for-grantedness of hegemonic spatialities.

The rise to prominence of socially constructed space, as produced rather than given, involves a transition in how space itself is theorized, what it means, how it is made. Harkening back to the famous seventeenth-century debate between Isaac Newton and Gottfried Leibniz, in which space assumed absolute versus relative properties, respectively, Barney Warf suggests that modernity portrayed space as a surface, i.e., as a flat, homogenous field in which differences could be minimized and places brought into an ocularcentric field of knowledge. Conversely, socially constructed, poststructuralist notions of space entail the metaphor of networks, such as the internet, which are forever partial, incomplete, and never fully known. Conceptually, the appreciation of networked space has taken various forms, such as globalized commodity chains, Massey's power-geometries, and Deleuzian rhizomes.

Political scientist Jeffrey Kopstein makes the intriguing argument that the collapse of communism across Eurasia in the 1990s opened a space for a laboratory in which the differential trajectories of multiple countries formerly dominated by (or part of) the Soviet Union could be compared and contrasted. Some moved happily into democratic societies, whereas others sank into abysmal dictatorships even worse than they suffered under their Soviet overlords. The geography of these transformations opened a spatial turn in political science, in which political change can be fruitfully seen as path-dependent and contingent rather than predetermined by any given set of institutional factors. In making this argument, he explores the provocative question as to whether geography is a meaningful category through which to explore such issues or whether space is only a proxy for other, aspatial causal forces.

Despite the earnest efforts of earlier approaches such as the Chicago School, Sociology has tended to ignore space and spatiality as constitutive features of social life in favor of perspectives that stress general explanatory laws and minimize the time-space contexts of social processes (e.g., modernization theory). Harry Dahms maintains that as sociologists have grappled with the enormous changes wrought by globalization, the field has come to appreciate space anew. In this perspective, space materializes at the intersections of deep, underlying forces and the surface appearances of everyday life and behavior. By incorporating space, sociologists both avoid the empiricism that continually confronts the field (and many others, including Geography) and are well positioned to appreciate the complexity of contemporary structural transformations and individual identity formation.

Space has played such a long and important role in literary criticism that it is difficult to know where to begin to summarize its significance. Pamela Gilbert's essay approaches the spatial turn in English literature by delving into sex and city, both the nineteenth-century version in London and its twentieth-century counterpart in New York. The city – that dense, complex, often bewildering environment that offers so many opportunities and possibilities for anonymity – has for centuries generated connotations of illicit sexuality, of adulterous affairs, casual encounters, prostitution, predatory attacks, and homosexuality. Drawing upon a wealth of tropes, poems, stories, novels, and television shows, Gilbert shows us that the spaces of the city simultaneously gave rise to particular forms of sexuality, legitimate or otherwise, and in turn were shaped by those practices.

Pragmatically and conceptually, the depiction of space has played a major role in the discourses of history. Santa Arias's essay contextualizes the relationship between geography and history by examining the role of space in the works by William Robertson and Juan Bautista Muñoz. She focuses on the place of cartography in historical accounts in order to explore eighteenth-century forms of geopolitics and show how the layering of history onto the map and the insistence on national contributions to geographical knowledge were turned into justifications of empire in the name of progress and civilization.

Colombian anthropologist Margarita Serge offers insights into the spatial turn in her discipline by delving ethnographically into the multiple ways in which people and landscape are intertwined in the region of Sierra Nevada de Santa

Marta. Over time, multiple groups shaped, and were in turn shaped by, this area, from Tairona to Kogui Indians, campesinos, guerrillas, and marijuana growers. The NGO, Fundación Pro Sierra Nevada, actively seeks to protect and preserve the region's cultural and biological diversity. In sharing her insights from this project, she reveals an Andean place that is constituted simultaneously in historical, material, and ideological terms, and how the status of being an "insider" or "outsider" is problematized through the act of understanding and navigating the locale's complex cultural and political dynamics.

Religion in all its varied forms has, of course, always been deeply spatial, an observation that geographical scholars of religious beliefs have long known. John Corrigan draws upon this lengthy corpus of scholarship to explicate how religious spaces operate at the interface of two worlds, the material and the theological. He notes that such places are always "polylocative," ever-fluctuating between the real and the imagined, the present and the hereafter. He traces this notion through a series of events and processes that are simultaneously temporal, spatial, secular, and religious, such as pilgrimage, migration, rituals, the inscription of the religious body, material culture (including food and music), religious practices, the state, and the construction of time and memory, all of which reflect, inform, sustain, and at times challenge the dominant norms of theological institutions.

For scholars concerned with colonialism, space has been an indispensable avenue through which one can appreciate the complex dynamics of the European conquest. In the case of Latin America, Mariselle Meléndez draws upon post-structural cultural geography and critical cartography to analyze how an eighteenth-century Peruvian newspaper, the *Mercurio Peruano*, was caught up within and amplified the political dynamics of the nascent independence movement, the construction of Creole identity, and the propagation of Enlightenment rationality in the New World. Such an exercise is useful in revealing how spatial discourses are not simply reflective of social and geographic transformations, but also constitutive of them.

Joan Ramon Resina explores the interface between film and geography, the realm in which the real and the cinematic bleed into one another. Film captures and reproduces space, bringing it to the eye and into consciousness in a manner no other technique can quite approximate. But what is gained and what is lost in this appropriation? Resina utilizes Luis Buñuel's *Land Without Bread,* a surrealist documentary about an impoverished region in Spain, Las Hurdes, to explicate how the real and the magical become blurred through the vision of the camera, a visual "surface of meaning."

From Kogui mythology to postcommunist Eurasia, from *Sex and the City* to Las Hurdes in Spain, does space retain any meaning as it has been embraced by other disciplines? As geographers opened doors to other fields – and went through them – space has acquired properties and qualities that few could have foreseen when they initiated the spatial turn. Far from the traditional, Cartesian notion of space as a set of physical places, contemporary thought throughout the social sciences and humanities reveals it as a variegated, complex, often bewildering series of different types of locations: physical, mythological, symbolic, imagined, linguistic, cartographic, perceptual, representational, i.e., space as suspended between matter and meaning. Indeed, as geographic thought has penetrated fields outside of geography, the nature of space has become ever more diverse.

Indeed, so wide-ranging are the types of space that have emerged in the wake of the spatial turn that geographers have lost control of their defining subject of study, the one topic that united a notoriously heterogeneous and schizophrenic discipline. All of these essays demonstrate, in some way, shape or form, that "space matters," not for the trivial and self-evident reason that everything occurs in space, but because *where* events unfold is integral to *how* they take shape. As historicism gradually loosens its tenacious hold over social thought, causality and context are inseparably fused. Space is not simply a passive reflection of social and cultural trends, but an active participant, i.e., geography is constitutive as well as representative. Those who relegate geography to epiphenomenal status do so at their own risk, depriving themselves of a critically important instrument with which to gain insight into the contingent logics of human affairs. By now this argument has been made clear a sufficiently large number of times that space has become indispensable across the social sciences and humanities. In this light, the spatial turn is irreversible.

2 Taking space personally

Edward W. Soja

There is no word for what I do, for what I passionately profess. I identified myself as a geographer at an early age and have accepted a series of qualifying labels over the years: regional, political, theoretical, development, quantitative, Africanist, critical, Marxist, structuralist, anti-humanist, neo-marxist, and more recently postmarxist, poststructuralist, postcolonial, postmodern. But geographer no longer seems enough, even with all its adjectival baggage, to describe someone who interprets the world by assertively foregrounding a spatial perspective. I put space first, before seeing things historically or socially, or as essentially political or economic or cultural, or shaped by class, race, gender, sexual preference; or screened through discourse, linguistics, psychoanalysis, Marxism, feminism, or any other specialized disposition. I try to see the world through all these perceptive lenses, but the primary focus is insistently spatial, informed, motivated, and inspired by a critical spatial perspective.

If geographer is not enough, what am I? If I foregrounded temporality rather than spatiality, naming would be easy. I would be called a historian, a transdisciplinary observer of lived time in all its multitude of expressions. Although there are deep parallels between the spatial-geographical and temporal-historical perspectives, however, there is no term like historian to describe the all-encompassing spatializer, the person who believes not just that space matters but that it is a vital existential force shaping our lives, an influential aspect of everything that ever was, is, or will be, a transdisciplinary way of looking at and interpreting the world that is as insightful and revealing as that of the historian. For me, what is true for the historical applies also to the spatial and I have devoted my academic career to advocating this critical comparability. Few others, even within the discipline of geography, are so assertively spatial, so convinced that spatial thinking is central to the production of knowledge and so driven by the need to inform others of the epistemological power of a critical spatial perspective.

What then are my nominal choices? Spatial theorist? Spatiophile? Spatiologist? Spatioanalyst? All these help, but are not quite there yet, for the comparison with the simple and self-confident identity of historian leads to further complications. Spatializing (or being a geographer) has never had the same intellectual prestige and recognized interpretive power as historicizing (or being a historian), even though there is nothing in Western philosophical and theoretical thought that unequivocally establishes an intrinsic privileging of time over space or history over geography. Yet such a privileging continues to prevail. So whatever we call the spatial equivalent of historian, that person must contend with an entrenched legacy of scholarly and popular discrimination and subordination maintained, even if out of conscious awareness, by the still hegemonic "history boys."[1]

Searching for a name is not just a playful exercise, for it opens up a distinctive viewpoint on the recent resurgence of interest in space, what is now being widely described as a Spatial Turn in the human sciences. It defines the Spatial Turn from the start as a response to a long-standing if often unperceived ontological and epistemological bias in all the human sciences, including such spatial disciplines as geography and architecture. This spatial advocacy is not against historical interpretation, an anti-history, nor is it a substitution of spatial for historical determination, as some skeptics have seen it to be. Reflecting the uneven development of historical versus spatial discourse, the Spatial Turn is fundamentally an attempt to develop a more creative and critically effective balancing of the spatial/geographical and the temporal/historical imaginations.

Over the past ten years, the number of dedicated spatial thinkers across all disciplines and modes of thought has been increasing exponentially, spreading a belief that the materialized and socially constructed spatiality of human life is just as revealingly significant, ontologically and epistemologically, as life's historicality and sociality (Soja 1996). Whether we are pondering the increasing intervention of the electronic media in our daily routines; seeking ways to act politically to reduce poverty, racism, sexual discrimination, and environmental degradation; trying to understand the multiplying geopolitical conflicts around the globe; or seeking new insights through academic research and writing, we are becoming increasingly aware that we are, and always have been, intrinsically spatial as well as temporal beings, active participants in the production and reproduction of the encompassing human geographies in which we live, as much and with similarly given constraints as we make our histories.

The Spatial Turn is still ongoing and has not yet reached into the mainstream of most academic disciplines. Its future expansion, however, has the potential to be one of the most significant intellectual and political developments of the twenty-first century. Framed by a series of autobiographical notes and by a belief that biographies are as much geographies as they are histories, what follows aims at gaining some further understanding of this still advancing and potentially epochal paradigm shift.

Geographical awakenings

Growing up enveloped in the heterogeneous densities of the Bronx helped make me a geographer by the time I was eight, and I have been a passionately committed proponent of thinking geographically ever since. My budding geographical imagination was locally grounded in the thick layers of social interaction and co-present cultural diversity that drenched the streetscapes of the outer borough. The street's densely socialized geography was intricately mapped out: marble games of various sorts in several specific locations; hit-the-penny, slug, and baseball card flipping on the square-lined cement sidewalk; hi-bounce rubber ball games played off certain stoops and curbs, with stickball taking up the entire street; and "four corners," the most popular game of all and my favorite, attracting multiple generations (and occasionally some heavy betting by the 'big guys') making almost constant use of the highly focal street corner. All of the children had detailed mental maps, varying with the seasons, of the use value of nearly every square inch of our lived streetspace.

From the start, however, my mental mapping always soared beyond the local, seeking to explore and vicariously experience the ever-widening other spaces that stretched outward from my street corner-defined homeland.[2] Everything on earth was alluringly present, just a map-read away or else at the end of a subway ride into the umbilical magic of hyper-connective Manhattan, the core to our periphery. My very being became quickly absorbed in thinking geographically, projecting myself into other spaces and places, imaginatively exploring faraway cities and regions and cultures as escape, education, adventure. I think I was born to spatialize, to celebrate my emplacement in that nesting of nodal regions that defines our being-in-the-world.

My geographical imagination was also and always particularly and intensively urban. Cities focused my attention. When I was 10 or so I was swept up by a compulsion to know, name, and locate every major city on earth. I compiled a list, hand-written at first and then typed out a year later, organized by country, with cities over one million written in red ink, those over 500,000 in black underlined in red, and all other cities over 100,000 just in black, making sure to include the capital city even if it were under my 100,000 threshold. I located them all in atlases and gazetteers, etching the political-territorial map of the world into my seemingly eidetic geographical memory.

The same urban excitement hit me again last year, when I came across a list prepared by the UN of the 430 or so global city regions of more than a million inhabitants, a new category of great contemporary interest to me. I spent hours restimulating my mental cartographies as I thought about an article I was writing on the urbanization of the globe and the globalization of the urban (Soja and Kanai 2008). As one might expect, China stood out with the largest number of million-plus cities, reflecting the most rapid urbanization-industrialization process the world had ever seen. With a quiet smile, I remembered writing a letter more than five decades ago addressed to the "Chinese Embassy, Washington, D.C." politely asking for a list of all Chinese cities over 100,000 to make my compilation complete, for none of my almanacs and gazetteers had long enough city lists for China. I never received an answer.

Finding Andorra

My formal education as a geographer started at about the same time. In my first school report at Public School 6 I shared my amazing discovery of Andorra, hidden away along the boundary between France and Spain.[3] I felt both wonder at its appearance, literally and figuratively, and a little dismay at my failure to know of its presence. I became determined to find out about all the other tiny countries hidden away from itinerant cartographic adventurers: Sark, San Marino, Liechtenstein, Monaco, Sikkim, Hunza. I drew maps of my unusual discoveries, such as the island of Samosir, with its Hindu temple, located in the middle of Lake Toba, itself located on the predominantly Muslim island of Sumatra, part of the Indonesian archipelago. Oh, how I wanted to be there, to take a trans-scalar journey, to do what today would be called Google-earthing in post-tsunami Aceh.

I kept trying to explain to my childhood friends the thrill that was attached to amassing geographical knowledge and vicariously visiting foreign places, but few would understand. I called myself a geographer but would not tell anyone else. Whenever asked what I wanted to be when I grew up, I never said geographer because I didn't know you could make a living doing what I was doing. No one else I ever met said they wanted to be a geographer. Who would pay you to do something that felt so good?

Geography was my secret life, full of imagined adventure and discovery in other worlds out beyond the familiar geographies I more directly shaped and was shaped by on and off the streets of the Bronx. Closer to home, I remember walking with friends to the archly dangerous (lots of bad guys hidden about) but magnetically attractive (especially for picking lilacs in April) Bronx Botanical Gardens. As we strolled along Southern Boulevard, we heard music and crowd noises some distance in from the main street, and when we followed our ears and nose (for the scents in the air were sumptuous) we found Italy! It was a hidden Italian neighborhood celebrating a saint's feast day on the streets, with food, dancing, pumping accordions, unaffected and unexpected joy.

I found out more about geography at Stuyvesant High School in deep and distant Manhattan, nearly an hour's subway journey from my home. Instead of woodworking or sheet metal shop, I was able to take what I was told was the only cartography class taught in any high school in the country. No high school taught geography as its own subject but I learned about map projections and how they always distort reality. Maps, I realized, can never present an accurate picture of the territory. Nevertheless, I learned how to print neatly with stencils and India ink pens, and I became a whiz with zipatone paste-on shading. The teacher appreciated my enthusiasm and told me that yes, Edward, it is possible to be a geographer. Never looking back or facing doubts about my future, or doing anything else to earn a living, I have been a practicing geographer ever since.

Into hypothetical worlds

New York City in the 1960s was a wasteland for advanced geographical education, the only affordable exception being Hunter College in the Bronx (now called Lehman), a public college a bus ride from home with a fully fledged Department of Geography and Geology. My formal education as a geographer would flow without interruption or diversion after four years at Hunter when, barely out of my teens, I left the Bronx behind to attend the University of Wisconsin.

I defined myself then—and now—as primarily a political geographer. Although I started with an interest in Central Asia, Sir Halford Mackinder's Heartland of the World Island, and the so-called geographical pivot of history (nice phrase that), and did my first seminar paper on new Chinese settlement in Sinkiang, I began shifting my regional interests in Madison to the resounding periphery of Africa and would become an active Africanist geographer over the next fifteen years. But strangely enough, what would affect me most during my one year at Wisconsin and trigger a lifelong interest in spatial theory and theorizing was a cartogram found in textbooks on climate, the egg-shaped 'hypothetical continent'.

This hypothetical and heuristic continent of climates offered a wondrous new way to look at the world, literally and figuratively. Stretching from pole to pole and bulging more broadly north of the equator to reflect the larger land masses in the northern hemisphere, this imagined supercontinent was marked by "expected" climate zones based on a conventional Köppen classification. It reflected an understanding of basic atmospheric dynamics, orographic effects, ocean currents, the shifting Inter-tropical Convergence Zone, all the general forces that shape climate in different parts of the world. In reading the map knowledgably, one could roughly predict temperature and rainfall patterns practically everywhere on earth: a remarkable condensation of geographical knowledge on one page. Mediterranean climates, with their distinctive pattern of dry hot summers and wet winters, were decidedly mapped on the west side of the hypothetical supercontinent in two latitudinal zones north and south of the equator. Looking at an actual map of the world's climatic zones, there they were lined up where they were predicted to be: the vast Mediterranean Sea region on the western edge of Eurasia, the Californian version at the same latitude in western North America, and even more amazing an equivalent zone in Chile and two others at the southern tips of Africa and Australia where they reached into the same southern hemisphere latitudes. One could also make good guesses about the landscape and vegetation in all these areas and enjoy the realization that these far-flung Mediterranean zones and their nearby cooler and wetter extensions were the source of nearly all the world's production of wine and olive oil. One could also find foggy coastal deserts (equatorward from the Mediterranean zones), identify areas most affected by monsoons or tornadoes, locate the warmest ocean currents, the tundra, the rainforests and savannas. The supercontinent did not exist. It was a creative figment of someone's geographical imagination, a synthetic mapping that stimulated a general understanding of an extraordinary variety of empirical real-world conditions. If this was what theory formation was about, then I must be not just a geographer but also a geographical theorist, seeking effectively evocative models of imaginary worlds that could not be found on the ground.

Seeking spatial theory

I followed up my absorptions with the hypothetical continent by taking a seminar with Glenn Trewartha on his book *The Earth's Problem Climates*, where he looked at the world's climate anomalies, places which did not fit the hypothetical schema. These demanded more detailed analysis to understand why they deviated from the norm. I remembered this later when, in the Ph.D. program at Syracuse University, I learned about mapping residuals from regression, identifying areas of geographical variation which the general regression model could not explain. I came to realize that what theories could not explain was just as important as what they could, as human geographies took on a multi-layered complexity that entwined the general and generalizable with the unique and particular, the nomothetic with the idiographic as they were called by the geosophers.

I left Wisconsin for Syracuse searching for more theoretical excitement in these hypothetical worlds, and found it with an odd combination of mentors, the brilliant young firebrand of the new quantitative and theoretical geography, Peter Gould (also an Africanist) and the more urbanely wise historical and cultural geographer Donald Meinig. My geographical imagination exploded with excitement and newfound zeal at Syracuse. My hypothetical worlds multiplied with the isotropic planes and spatial models of the new geography as taught by the irrepressible Gould; while, in a very different, more traditionally geographic mode, Meinig invigorated my thinking about cultures and culture regions and the historical geography of world civilizations through his own vividly descriptive modeling of culture hearths, cores, and peripheries.

My geographical imagination was particularly stimulated by Peter Haggett's *Locational Analysis in Human Geography* (1965). Never was geography more enchanting, more alluring. There were not just the familiar surface appearances, the mapped material spaces, the areal variations that could explain other areal variations (the bread and butter of statistical geography), there were other worlds, at least five formative layers building on one another to create and tie together the vividly spatial organization of human society: movement–networks–nodes–hierarchies–surfaces. Everything was there: Hagerstrand's propagation of innovation waves, gravity models of human interaction, Von Thunen rings of concentricity, Alfred (not Max) Weber's industrial triangles, location quotients, Loschian lattices and Christaller's central place theory, regionalization methods, the brim-full space economy, everything taking shape through adjustments to the friction of distance rather than any physical environmental factor. All this would be there, they told me, even if the earth's surface were as smooth as a billiard ball. Imagine!

I fed off Haggett and what I felt were the best possible teachers of the new and the old geography, Gould and Meinig, and became an even more ardent missionary for geographical thinking, armed with new tools to spread the exciting word to all those who were as yet unaware. I was no longer just a geographer. Geography became the spatial organization of human society, what I would later call (pace Lefebvre) the spatiality of social life, and I was its analyst and theorist, charged with convincing others of the powers of spatial analysis and explanation. My evangelical spatializing as a doctoral student was aimed primarily at political scientists, especially those in comparative politics and interested in the independence movements and nation-building efforts that were spreading throughout

Africa. Challenged by Gould to revolutionize the moribund field of political geography, I boldly marched off to Africa in 1963, stopping for a few months at the School of Oriental and African Studies in London to learn Swahili.

Mathematical intermission

In 1965, after returning from eighteen months researching the geography of modernization in Kenya for my dissertation, I began my first teaching job in the now defunct Department of Geography at Northwestern University. In the next two years, I was married, had my first of two children, completed writing my dissertation, and was carried forward by the compelling flow that had always been shaping my life to become a geographer by profession as well as persuasion.

Geography at Northwestern was swept up by geographical mathematics and statistics at an absurdly high level of abstraction. Our best doctoral students followed a unique program that, in addition to advanced courses in calculus and linear programming, moved them through a sequence of so-called geography courses beginning with explorations of point patterns in the plane (fitting probability functions into distributions of dots of various sorts, from county seats in Iowa to freckles on your arm), moving on to the next dimension in linear patterns in the plane (with all sorts of network models and connectivity measures), and finally to the ultimate in n-dimensional space exploration, trend surface analysis, in which one could exactly reproduce any given geography through a series of statistically defined surfaces of accumulating complexity, each one absorbing variance until all was soaked up in the end. Haggett's model was embedded therein, but it was hammered into inert equations and geometrics.[4]

I stayed at Northwestern for seven years, including two spent in Africa. For an enthusiastic geographer, Northwestern was a peculiar interlude. I published enough books and articles on Africa to get tenured before I was 30. I also published my first serious attempt to retheorize political geography, an AAG Resource Paper on *The Political Organization of Space* (1971), in which I explored human territoriality as central to the study of comparative politics. But there was little to re-excite my geographical imagination. I needed a change and in 1972 I joined the Urban Planning Department at UCLA and have been there ever since.

The spatial turn begins in Paris

The 1960s saw explosive urban unrest spread around the world and from the rubble and ashes grew a revolutionary new way to think about space and the powerful effects of specifically urban spatiality on human behavior and societal development. I missed this largely Parisian revolution in spatial thinking, having been primarily absorbed in and diverted by the so-called and primarily Anglo-American quantitative and theoretical revolution in geography. What was happening in Paris, however, was much deeper and broader. A new way of thinking about space and its forceful effects was beginning to take shape, reversing a hundred years of historicist hegemony in which specifically urban spatial causality, what would much later be called the stimulus of urban agglomeration, was buried under the privileging of historical and social theoretical discourses.

Up until the late 1960s, the only significant academic movement promoting a forceful spatial perspective was the Chicago School of Urban Ecology. For a brief moment in the interwar years, a band of determined spatializers centered in the departments of sociology and geography at the University of Chicago created a collective intellectual community aimed at exploring how human behavior is directly shaped by the urban environment. Crudely activating and empowering urban spatial causality via ecological processes, they saw the city generically as the "embodiment of the real nature of human nature . . . an expression of mankind in general and specifically of the social relations generated by territoriality" (Janowitz 1967). For most social scientists this ecological spatialization went too far and after World War II the influence of the Chicago School deteriorated, kept skeletally alive only in narrow strands of urban economics and urban geography. What happened in the late 1960s in Paris was therefore highly unexpected, nearly incomprehensible, and thoroughly easy to ignore, at least in the English-speaking world.

The two major figures in this peremptory and soon deflected initiation of the Spatial Turn were Henri Lefebvre and Michel Foucault. We know very little about how they influenced each other if at all – several books and dissertations are needed on this subject – but their co-presence in Paris at this unsettled time is undeniable and so too is their nearly simultaneous development of essentially similar ideas about the ontological significance of space and the powerful forces that emanate from the spatiality of human life (Soja 1989, 1996). The coincidences, literally and figuratively, are remarkable.

The French intellectual tradition has always had a greater spatial sensibility than the Anglo-Germanic, from the *philosophes* (Montesquieu, Rousseau, Voltaire) of the eighteenth century to the twentieth-century surrealists, situationists, and historians of the Annales school (e.g. Braudel), and there were many thinking and writing about space, geography, and urbanism in 1960s Paris. But only Lefebvre and Foucault explicitly began a radical rethinking of the ontological, epistemological, and theoretical relations between space and time and, by implication rather than direct discussion, geography and history. From the contemporary perspective of the Spatial Turn, one can read into their writings a series of reconfigurative arguments.

Retheorizing space

The foundational moment for what would eventually become the Spatial Turn was an assertion of the ontological parity of space and time, that each was formative of the other at a most basic existential level, with neither being intrinsically privileged. Foucault, uncharacteristically, was most explicit about this in his lecture notes for a radio presentation published as *Des Espace Autres* or "Of Other Spaces" (Foucault 1986). After noting earlier tendencies in French philosophy via Bergson to privilege history, he asked almost sarcastically why is it that so many think of time as movement, dynamic, dialectic, development, process, while space is considered fixed, dead, unproblematic background, the stage or container of social processes and history. In this view, social causality and the force of human will was carried only by time and history, with space and geography merely reflecting the social drama, the essentially historical narrative.

Rejecting this privileging of time over space, Foucault assertively placed the origins of this obsession with history, as he called it, in the second half of the nineteenth century, in what retrospectively can be described as a far-reaching ontological distortion of Western social thought. This subordination of space and spatial thinking was carried forward to the present via an engrained social historicism that has shaped all the social sciences as well as scientific socialism, Marx's essentialist historical materialism. What was needed, then, was a re-empowerment and rebalancing of the spatial-geographical with regard to this privileged social if not socialist historicism, without denying the latter's extraordinary interpretive insights.[5] Both Lefebvre and Foucault would seek this counter-balancing reassertion of space through all their writings from the late 1960s to their deaths.

Both of these ardent spatializers would also develop the idea that existing modes of spatial thinking (even in France) were not adequate to the task of creating ontological and epistemological parity between space and time, geography and history. Hence, each set about constructing a more comprehensive and enriched alternative mode of thinking about space that would build upon the strengths of existing spatial modes of thought but also move beyond them to open up new possibilities for knowledge formation. Something like this had already happened in physics, with the triad of space-time-matter/energy; now it was time to do something similar, analogous if not homologous, in the human sciences, a triple dialectic of the temporal-historical, the spatial-geographical, and the social-sociological imaginations.

Foucault called his 'other' way of thinking about space heterotopology and encouraged looking at every created space, from the most intimate spaces of the body and the home, to the global spaces of geopolitics and military conflict, as heterotopias, redefining this term to exemplify his different or heterotopological mode of thinking about all spaces.[6] Lefebvre opened up his alternative through his conceptualization of what he variously called lived space (*espace vécu*), spaces of representation, or representational spaces. Whatever English term is used, the task was to open up a new perspective, one that would be significantly different from what conventionally existed.

Both shared the view that spatial thinking in the past has been fundamentally and, to some degree, restrictively, bicameral. Spatial discourse was dominated by two alternative modes, one emphasizing material conditions, mappable spatial forms, things in space (Lefebvre's perceived space or spatial practices) and the other defined by mental or ideational imagery, representations, thoughts about space (Lefebvre's conceived space). Both would also praise the best work using this bicameral duality and urge continued application, but at the same time Foucault and especially Lefebvre would discuss the critical limitations of each mode.

Emphasizing the empirical expressions of spatial practices, that is, describing and interpreting material geographies, dominated traditional geographical analysis and produced significant and useful factual knowledge about the objective, real world. But there was a tendency to fixate on materialized surface appearances and directly measurable patternings, creating an illusion of opaqueness that could block deeper understanding of the causal forces underpinning these surface expressions. Similarly, another illusion arose from focusing too emphatically on representations or images of material spaces, on idealized visions of the world. Here the material geographies became translucent, lost in the luminous search for deep structures of causality as the imagined took precedence over the real. Again, brilliant achievement was possible through this representational mode of spatial thinking, with both Foucault and Lefebvre noting in particular the work of Gaston Bachelard on *The Poetics of Space* (1969), but these representations were not enough.

What was needed to rebalance spatiality with historicality and sociality was another, more comprehensive and combinatorial mode of spatial thinking, one that built upon the traditional dualities (material-mental, objective-subjective, empirical-conceptual) but moved the search for practical knowledge beyond their confines to open new ground to explore. Reacting politically in different ways to the events of May 1968 in Paris, Lefebvre and Foucault saw potential emancipatory power in a consciously spatial praxis based in a practical and political awareness that the geographies we have produced (or were produced for us) can negatively affect our lives but that we can act to change these unjust and oppressive geographies. Also clear to both was the necessity to lead spatial thinking in a new direction, for the old ways had too many dead ends.

These creative intimations of a radical Spatial Turn, demanding a paradigm shift of wide-reaching transdisciplinary proportions, were not allowed to have a lasting effect. Emphasizing urban spatial causality was simply unacceptable for most Marxists, including the Marxist geographers and sociologists (David Harvey and Manuel Castells in particular) who were achieving growing influence in urban studies in the 1970s. Foucault was simply ignored for his presumed de-radicalized personal politics, while Lefebvre, recognized as one of the twentieth century's leading Marxist philosophers and urban thinkers, was praised for his insights but ultimately set aside for promoting urban spatial causality, even urban revolution, too intensely.[7] In the wake of these deflections, the Spatial Turn would remain undeveloped if not dormant for more than two decades.

Geographer in exile

I moved into Urban Planning at UCLA without realizing the professional and perceptual repercussions until much later. From the perspective of American geography as a discipline, my move seemed to indicate that I had turned in my identity as a true geographer. I remained a distant cousin perhaps, but was no longer part of the nuclear family. The move, however, was extraordinarily stimulating for me as a spatial theorist and advocate. Urban planning was not as tightly disciplined and intellectually introverted as geography. I had greater freedom and encouragement than I would ever have in a department of geography to expand my theoretical horizons and promote geographical ways of thinking—as long as I was willing to address at least some of the immediate needs of professional and practicing planners. Indeed, the pressure to make theory practical, after some defensive waffling about the pureness of theory, sharpened my geographical imagination.

At first, I immersed myself in the new Marxist geography and abandoned as ideologically diversionary distractions nearly all of the other approaches I had learned as a graduate student and young professor. I started anew, reborn not just as a geographer but as a specifically and ardently Marxist geographer-cum-planner. This raised my spatial advocacy to an even higher pitch, as I now thought I could not only comprehend descriptively the human geographies that blanketed the surface of the earth but get underneath them to find the social processes and social relations of production and capitalist accumulation that were profoundly shaping spatial form. But why, I asked myself, weren't my radical comrades as enthusiastic about the stimulating powers of urban spatial causality, how spatial process shape social form?

I came out again as a critical spatial theorist after a decade of rethinking and reading Lefebvre with the publication of "The Socio-spatial Dialectic" (Soja 1980). In what was originally titled "Topian Marxism," I expressed my disappointment with Marxist geography's fealty to tradition and consequent failure to adequately explore the compelling power of the critical spatial imagination. The key argument, drawing heavily on Lefebvre, was that spatial processes shaped social form just as much as social processes shaped spatial form. It seemed so obvious to me that spatial relations of uneven development were just as important in theory and political practice as social relations of class.

At first, I aimed my critique mainly at my fellow Marxist geographers, nearly all of whom at the time seemed unnecessarily wary, almost fearful, of giving explanatory power to anything other than social class, leading many to label me a spatial fetishist (albeit with deluded good intentions). I persisted nonetheless. Society is formatively spatialized from the start, I asserted, in much the same way as space is formatively socialized. Socialization and spatialization were intricately intertwined, interdependent and often in conflict. Neither the spatial nor the social should be privileged over the other, especially in the historical-geographical materialism being so insightfully created by David Harvey and a few others.

Building a much broader critique of all variants of modern geography, from the old positivism to the new Marxist, feminist, and cultural versions, I repeated my arguments more substantively in *Postmodern Geographies* (Soja 1989). In what I continue to consider my most important contribution to the critique of modern geography and to the Spatial Turn, I connected the socio-spatial dialectic to its philosophical twin, the mutually constructive interplay of history and geography, a spatio-temporal dialectic that, I argued, became ontologically distorted in the late nineteenth century (following Foucault) and has persisted to the present.

Formulating this triple dialectic, or trialectic as I would later call it, and noting the persistently prioritized power of the socio-historical pairing over the socio-spatial and spatio-temporal, gave me new insight into why geography and critical spatial thinking had been so relatively and relationally neglected, so peripheralized in the academic and intellectual division of labor for at least the past century. Thinking spatially, the geographical imagination, and geography as a discipline had been effectively buried under a space-smothering social historicism, an epistemological occlusion activated by a peculiar privileging of the social and historical over the spatial that continues to shape contemporary social thought. Modern geography itself was also to blame, as it failed to see its internal restrictions in scope, ignoring the powerful critiques of Foucault and Lefebvre.

Nearly everything I have written since 1989, especially including *Thirdspace* (1996) and *Postmetropolis* (2000), has been aimed at making essentially the same argument that was at the core of *Postmodern Geographies.* Over and over for nearly 20 years, I have been trying in many different ways to convince others of the extraordinary power of thinking spatially, using the socio-spatial dialectic to see not only how social processes shape and explain geographies but even more so how geographies shape and explain social processes and social action. The trialectic of spatiality–historicality–sociality and the associated recognition of urban spatial causality brought back into the contemporary discourse what Lefebvre and Foucault had called for in the late 1960s. For the present, strategically putting space first as an interpretive framework was necessary to address the nineteenth-century ontological distortion, combat the enduring force of social historicism, and introduce a new and different mode of critical spatial thinking and praxis.

As I would discover much later, a younger generation of geographers received some inspiration from my work to explore beyond the conventional confines of modernist geography, tired of the titanic internal squabbles that were raging in the 1980s between hegemonic positivism and rebellious Marxism, humanism, and feminism. For many more established senior geographers, however, my enthusiastic advocacy of a new mode of spatial thinking was met with suspicion if not anger. Unusually harsh reviews by geographers made me realize that I was being perceived as creating my own personal version of geography, different than that which occupied the vast majority of geographers; and that by trumpeting its virtues I was simultaneously and perhaps solipsistically insulting these established geographers for not being smart enough to explore my preferred way of thinking geographically. For those outside the disciplined cocoon of geography, however, it was a different matter. By the mid-1990s, I had become an influential reference point for the growing number of newcomers to critical spatial thinking.

The postmodernization of geography in Los Angeles

To those who saw my writings after 1980 as marking a radical shift in perspective, I would respond that I merely moved from studying the geography of modernization (and the modernization of geography) in Kenya to investigating the geography of postmodernization (and the postmodernization of geography) in Los Angeles. What remained essentially the same was the driving force of my spatial advocacy. My writings on the astonishing transformations of the geography of Los Angeles gave invigorating support and a more compelling power to my theoretical critique.

I do not intend to rehash the debates on postmodernism in geography here, nor will I try to defend the idea of a Los Angeles "school" of critical urban studies.[8] What needs to be said, however, is that over the 35 years that I have been a professor of urban planning at UCLA, I have been unusually fortunate to be exposed to a truly remarkable agglomeration of creative spatial thinkers. Whatever I may have achieved arises from being part of this incomparable group of scholars, a partial listing of which would include Allen Scott, John Friedmann, Michael Storper, Margaret FitzSimmons, Mike Davis, Dolores Hayden, Susan Christopherson, Nicholas Entrikin, John Agnew, Costis Hadjimichalis, Dina Vaiou, Goetz Wolff, Marco Cenzatti, Margaret Crawford, Paul Ong, Clyde Woods, Charles Jencks, William A.V. Clark, Susan Ruddick, Barbara Hooper, Mustafa Dikec, Olivier Kramsch, Ferrucio Trabalzi, Julie-Anne Boudreau, to name just those affiliated in some way with urban planning and geography at UCLA.

Los Angeles proved to be an extraordinary laboratory for the production of spatial theory, for exploring urban spatial causality in its many expressions, and for demonstrating the insightful power of a critical spatial perspective. Los Angeles-based research and writing played an important role in stimulating the Spatial Turn and spreading spatial thinking across a growing number of subject areas. Perhaps inevitably, the body of work emanating from the spatial thinkers of Los Angeles has also generated criticism, especially from those who perceived its enthusiastic assertiveness as the imposition of a LA-specific model and mode of thinking in places where it does not fit, or as arrogantly defining an exclusive and self-referential club (Gottdiener 2002). In balance, however, there is some justification in concluding that if the Spatial Turn began in Paris, only to fade from view, it arose again with renewed force in Los Angeles.

Thinking spatially about Los Angeles, while based on rich empirical detail, was intentionally more nomothetic, aimed at producing generalizable knowledge, than was the case for most urban research. The aim was not to show the unparalleled uniqueness of the city but rather how localized knowledge can help to understand what is happening in other cities around the world. Los Angeles served as a laboratory, a hypothetical continent, for developing new urban theories focused on the restructuring processes that have been reshaping cities everywhere in the past forty years, in particular the formation of a new (post-Fordist, flexible, information-based) economy, the globalization of capital, labor, and culture, and the development of new information and communications technologies. This work added to what was already a key feature of the Spatial Turn, its overlap with a resurgence of interest in cities and regions. For many around the world and in practically every discipline, the Spatial Turn has also been a turn to urban and/or regional studies as well.

Beyond geography: the widening scope of the spatial turn

As a cumulative and fast-flowing diffusion of a spatial perspective across nearly every discipline, the Spatial Turn burst onto the academic scene some time in the mid-1990s. Viewed from the standpoint of the traditionally spatial disciplines, geography and architecture in particular, this spreading spatial advocacy was more of a heterogeneous and eclectic expansion than an actual redirection of geographical thought and spatial theory. What was once confined primarily to the core of these disciplines was expanding into an almost untouched periphery, much like the spreading urbanization of suburbia and the rise of new industrial spaces that was transforming the modern metropolis in the last decades of the twentieth century (Soja 2000). Beyond the boundaries of geography, where relatively aspatial intellectual landscapes prevailed, creative new clusters of spatial thinking and interpretation started to form. A spatially lifeless periphery was becoming actively spatialized, rather superficially in some areas but much more deeply energizing in others.

As spatial thinking began to flourish outside geography, most geographers remained relatively unaware or indifferent. The few that cared welcomed their newly spatialized colleagues, but often assigned them a subordinate role in the intellectual division of geographical labor. They were responsible for metaphorical and aesthetic assembly, for example, but design and creativity were kept at home, in the inner sanctums of the newly formed but loosely defined subfield of critical human geography. Most attempts to break open the disciplinary cocoon to alternative spatial perspectives were typically deflected away or summarily dismissed, as probably the majority of geographers reacted to the Spatial Turn by defending their traditional modernist turf.

The defensive disciplinary response often involved a vigorous rejection of the postmodernist critiques that were so closely associated with the Spatial Turn. Over the past ten years, the positivist and descriptive core of geographical analysis has refortified its centrality, sustained in part by large flows of financial support for the advancement of Geographical Information Systems (GIS). At the same time, the older critical periphery, consisting of only occasionally intersecting strands of relatively orthodox Marxism, feminism, and deeply rooted cultural geography, has been struggling to retain its former spheres of influence in a now dramatically expanded academic world of geographical inquiry.

The critical fringe has been far from silent in the development of critical spatial theory. While many were defensive and inward-looking, others such as David Harvey, Doreen Massey, Derek Gregory, Gillian Rose, Neil Smith, and Denis Cosgrove reached out to engage with a growing range of disciplines, effectively diffusing critical spatial perspectives in many different directions. After the mid-1990s, however, leadership began to shift to a generation of critical geographers who moved beyond their Marxist, feminist, or cultural geography roots to explore new ground through direct engagement with such fields as economics, anthropology, psychoanalysis, film studies, literary criticism, and international relations. Through a process of hybridization, it has become increasingly difficult today to draw boundaries between who is a geographer and who is not, for the unprecedented transdisciplinarity of the Spatial Turn is making almost every scholar a geographer to some degree, in much the same way that every scholar is to some degree a historian.

It is probably safe to say, however, that the Spatial Turn today, while drawing on a significant number of scholars trained in geography, has been having its deepest effect outside the discipline. This reconfigurative diffusion of spatial thinking has progressed in stages, advancing with great unevenness in the social sciences[9] but with unusually broad effect across the humanities. The initial stage of diffusion has been simply additive, involving the widespread use of spatial terminology and metaphors such as mapping, regions, place, space, territory, location, geography, cartography to suggest at least a dimensional spatiality to whatever subject is being discussed. I would think that every discipline in the human as well as the physical and medical sciences (regions of the brain, atlases of human anatomy) has in one way or another experienced at least this level of spatialization.[10]

This widespread adoption of spatial terms is not insignificant, for it is indicative of a growing change in awareness of and familiarity with the relevance of space and geography. It has been assisted by the new technologies of informatics, multi-media communications, and the internet, especially given the persistence of such encompassing and evocative terms as cyberspace and the widespread use of Global Positioning Systems, GIS, and Google Earth. The growing attention to economic and environmental globalization processes has also been an important factor in spreading a broad overlay of spatial discourse. It is no longer such a peculiar surprise to see a geographer join the "history boys" and other recognized pundits to give perspective to the television news or comment on major current events.

In some areas, spatialization has sparked the development of growing research and teaching clusters specializing in spatial thinking and analysis. Plugging "spatial turn" into a computer's search engine brings up the following fields where the specific term is being used: cultural studies, cultural theory, critical theory, literature, film, history of science, history, environmental studies, organizational theory, media studies, comparative education, philosophy, historical sociology, sociological analysis, poetry, Hispanic literature, religion, theology, Bible studies, planning practice, history and sociology of education, accounting, philosophy.

The Spatial Turn is seen by many as following other "turns" such as the linguistic, cultural, and postmodern. However such sequencing is defined, the Spatial Turn has evolved with close connections to critical cultural studies, with particular attention given to such spatially insightful postcolonial cultural critics as Edward Said, Gayatri Spivak, Homi Bhabha, and Arjun Appadurai. The cultural studies connections have also helped to extend critical spatial thinking into anthropology and the new ethnography, feminist and anti-racist critiques, media and communications studies, and comparative literature and literary criticism, including such specialized areas as Latina novel-writing and poetry. (For more on these fields of critical spatial discourse, see other chapters in this volume as well as n. 10.)

A recent publication (Brauch *et al.* 2008) illustrates the extraordinary effects of the Spatial Turn in a highly specialized subfield. *Jewish Topographies: Visions of Space, Traditions of Place* is the product of an interdisciplinary research group at the University of Potsdam, Germany, which over its life has produced about 30 dissertations and numerous articles, books, and exhibitions. The book explicitly builds on critiques of historicism, Lefebvre's triad of perceived–conceived–lived spaces, and other core ideas that have impelled the Spatial Turn everywhere.

There is also a list of "Conferences on Jewish Space" that have been taking place around the world since 1990, the most recent of which have been several in Potsdam and others at Stanford University (2003), Penn State (2004), University of Cape Town (2005 and 2007), Vilnius, Lithuania (2006), Lehigh University (2007), Karl-Franzens University in Graz, Austria (2007), and the International Cultural Centre in Krakow, Poland (2007).

Each of these more specialized spatial turns has its own geohistory that needs to be written, but I will take a closer look here at four disciplinary clusters where, from my own personal experience, the Spatial Turn has been particularly interesting and creative.

Thinking spatially in the visual arts

Art historians, critics, and practicing artists were among the first to respond positively to *Postmodern Geographies* after its publication in 1989. Some geographers saw my arguments as deeply flawed by an arrogant if not authoritarian masculinism that was amplified even further in David Harvey's *The Condition of Postmodernity,* published in the same year (Massey 1991). This emerging feminist critique included a particularly cutting attack on "men in space" by Rosalyn Deutsche (1996), an established art historian and critic who had already begun to explore the various new geographies.[11] While the feminist critiques deflected attention from *Postmodern Geographies* within geography, especially in England, Deutsche's writings may have given added interest to it in the theory and practice of the visual arts.

Well before these critiques, however, several art scholars went right to the heart of my critique of historicism and the ontological distortion that privileged time over space, history over geography, since the mid-nineteenth century. I learned from them that their field had experienced a mini-spatial turn in the interwar years, inspired largely by Walter Benjamin. Benjamin and others recognized the degree to which the critical power of the historical narrative could choke off an appropriate appreciation for visual and spatial representation in art. Benjamin effectively "spatialized time," cracking open the narrative's stranglehold to allow the visual and spatial to flourish (Gregory 1994: 234). These debates, however, were never formally theorized by art scholars and instead entered almost subliminally into the disciplinary culture. I was seen as bringing these ideas to the surface again and presenting a more detailed and philosophical theorization, linking the reassertion of space to structuralist (including Marxist) critiques, probing beneath surface appearances to see the social class and other personal and political forces shaping the creative impulse.

Another line of connection came from my discussion of John Berger, one of the world's leading art and literary critics and a creative specialist in what he has called "ways of seeing." Berger's writings (1972, 1974, 1984) have been especially sensitive to the reassertion of space versus time in the arts and literature and he can be seen, along with Lefebvre and Foucault, as an early purveyor of the Spatial Turn. His critique of social historicism is brilliantly conveyed in the following passage, where he moves from the declining symbolic power of portrait painting to the crisis of the modern novel.

> We hear a lot about the crisis of the modern novel. What this involves, fundamentally, is a change in the mode of narration. It is scarcely any longer possible to tell a straight story sequentially unfolding in time. And this is because we are too aware of what is continually traversing the storyline laterally. That is to say, instead of being aware of a point as an infinitely small part of a straight line, we are aware of it as an infinitely small part of an infinite number of lines, as the centre of a star of lines. Such awareness is the result of our constantly having to take into account the simultaneity and extension of events and possibilities . . . Prophesy now involves a geographical rather than historical projection; it is space not time that hides consequences from us . . . Any contemporary narrative that ignores the urgency of this dimension is incomplete and acquires the oversimplified character of a fable.
>
> (Berger 1974: 40)

My work reached an expanded audience of artists and urban activists after an appearance as one of the 100 Guests presenting lectures over the 100 Days of Documenta X in 1997, the world's largest art exhibition, held every four years in Kassel, Germany. Documenta X, directed by the French curator Catherine David, was aimed at reviving art as a political instrument and strategy of urban protest. It was also rooted in a reassertion of the spatial writings of Foucault and Lefebvre and successfully stimulated a significant expansion of interest in urban spatial causality and politics among visual and performance artists.

Ever since 1997, I have been invited by many groups of arts scholars and practitioners to lecture specifically on the Spatial Turn and related topics. These include a keynote address on "The City as a Vehicle for Visual Representation" at the annual meeting of the International Association of Art Critics held at the Tate Modern in London in 2000; a lecture and workshop at Arteleku, the leading Catalan art center and museum in San Sebastian, Spain, in the same year; presentations at the Museum of Modern Art in Los Angeles (2002), the Museum of Contemporary Art in Belgrade, Serbia (2006), the Andalusian Center for Contemporary Art in Seville, Spain (2006), the Kampnagel Art Center in Hamburg, Germany (2007); and a keynote address at a conference on "Sense of Place" organized by Site-ations International and the European Artists Network in Sligo, Ireland (2006).[12]

The Spatial Turn, especially in tune with urban activism, seems to be particularly well established in the visual arts and is expanding in innovative directions, with spillovers taking place in music, dance and choreography (the Greek term *choros* having close connections to space and geography), performance art (Gómez-Peña 1993), and other areas. An especially fertile arena for creative injections of a critical spatial perspective has been film studies, where the space-time relation and the links between cinema and the city have received particular attention (Bruno 2007; Sobchak 1997; Clarke 1997; Al Sayyad 2006). Both the visual arts and film studies are likely to expand significantly in the future as crossroads for productive interaction between new and old geographers.

God, space, and the city: the spatial turn in theology

I discovered several years ago at a conference at the Chicago Divinity School where I was asked to speak about the Spatial Turn and critical spatial theory as it relates to the reconceptualization of Nature that my work had become very popular in the esoteric field of eschatology, the study of end-states such as heaven and hell (Soja on "Seeing Nature Spatially" in Albertson and King 2008). I had become aware earlier that my work was being widely used in Bible studies (Flanagan 1999, 2001; Kohn 2006), but eschatology? What was going on here? It was explained to me that eschatologists were using *Thirdspace* (1996) and my concepts of critical thirding-as-othering and trialectics as a means of opening up new ways to explore the multiplicity of dimensions through which heaven and hell could be described as putative lived spaces. The Spatial Turn had seemingly reached its outer limits.

My inquiries led to an invitation to a conference on "God/City/Place: Interdisciplinary Perspectives," held at the Lincoln Theological Institute of the University of Manchester in 2006. There I addressed a plenary audience of critical and political theologians and commented on a Swedish scholar's dissertation on eschatological spaces. I learned rather quickly that theology, as much as any other field, specializes in rhetoric, the art of developing convincing arguments, and in order to hone these rhetorical skills theologians closely follow developments in the field of critical and cultural studies. For many years now, significant attention has been given in various subfields of theology to critical spatial theory and progressive if not radical urban spatial politics.

In "Theology in its Spatial Turn: Space, Place and Built Environmental Challenging and Changing the Images of God," Sigurd Bergmann (2007), of the Department of Archeology and Religious Studies, Norwegian University of Science and Technology, summarizes and explains theology's forays into spatial thinking. He writes that "Theology's reflections about space and place provide a deep challenge and an urgent necessity for theology to become aware of its embeddedness in the existential spatiality of life." More specifically, he goes on to discuss

> the space of Creation, God's spatiality, the loss of place, and spiritual practices as well as justice, redemption and aesth/ethics [sic] in the built environment and the ecological city . . . [and also] God in the city, lived religion, architecture, eco-feminism, geography and religion, land, new religiosities, pilgrimage, movement and mobility.

Drawing heavily on *Thirdspace* is "The Trialectics of Biblical Studies" (Flanagan 2001), presented first as a presidential address to a biblical society and presented again to a meeting of archeologists at a "Construction of Ancient Space Seminar," where many different scholars explored the relevance of spatial perspectives to research on the ancient world and especially debates on the origins and development of cities (Berquist 2002). The Spatial Turn in theology is interesting in itself for its breadth, depth, and seriousness, but it also serves as a bridge into another specialized realm where the Spatial Turn has even deeper roots and has progressed even further.

Archeology and urban spatial causality

No other discipline has been as closely tied to geography and the development of spatial theory in all its forms as archeology. I remember when I was a student at Syracuse being told that central place theory was being used by archeologists (somewhere in Turkey, if I remember correctly) to project the likely location of the largest trading centers in a network of urban settlements in which only a few smaller sites had been found. It was exciting to discover how useful analyses of point patterns on the plane could be. I would learn later that a new subfield called processual archeology was built heavily on the spatial science being developed in geography starting in the 1960s.[13]

Particularly entwined with developments in geography was the leading archeological theoretician, Ian Hodder, whose father I had known as a prominent Africanist. Hodder's theoretical archeology evolved in close association with developments in spatial theory, from the quantitative geography and location theories of positivist geography, through the three main strands of the Marxist-structuralist, feminist, and cultural-phenomenological critiques of positivist geography, to the (post-processual) eclecticism that emerged from more recent postmodern, postcolonial, and poststructuralist critiques. His approach has always been carefully selective and rooted in the practical needs of archeological investigation, but Hodder has orchestrated an unusually intricate and intimate connection between geography and archeology that is unparalleled in any other discipline, a somewhat ironic engagement given the inherent attractions of archeology to historicism.

I discovered Hodder's spatializing most revealingly through a nearly 9,000-year-old wall painting that would also open up for me surprising connections to a wider field of contemporary geographical economics. Let me explore this panoramic urban mural further, for it illustrates the extraordinary insights being generated by the Spatial Turn in our understanding of both the past and present. In the late 1990s I worked with the very spatial sociologist and urban planner Janet Abu-Lughod (who I met back in my Northwestern days) on a joint project for the Getty Trust on the Arts of Citybuilding. We were hoping to connect the critical traditions of art and urban studies in creative new ways, building on the connections between art and spatial theory mentioned earlier. As part of my work on the project, which was never completed, I looked at how the city and urban landscape of Los Angeles had been depicted in paintings in the twentieth century.

My research and writing on Painting Los Angeles was never published, but it led me to one of the most exciting discoveries in my geographical life. I had been thinking for some time that perhaps all creative art is in one way or another urban art, generated in the dynamic crucible of urban life, the spatial specificity of urbanism. I explored the voluminous classic texts on art history (those history boys again) looking for what was considered urban art, but could find no mention whatsoever of anything resembling urban spatial causality. The Spatial Turn in art history had not yet reached the textbooks. But there seemed to be some agreement among the texts on what is considered to be the oldest known "landscape" painting: a mural painted in bold colors on two walls of a house in a settlement called Çatalhöyük in south-central Anatolia around 6500 BC.

The wall painting consisted of dozens of what looked like the footprints of housing compounds strung together beneath an apparently erupting cinnabar-colored volcano. There was a suggestion in the *Guinness Book of World Records* that this was the world's first nature painting, but it was neither a nature painting nor a true landscape. It was a cityscape, the earliest known painting of a permanent urban settlement, the first intentionally built environment. The mural captured the movement from the raw to the cooked, pristine to transformed Nature, nomadic hunting and gathering to a settled and sedentary life in one location, in other words the beginnings of the urbanization process. It depicted not animals and hunters or colorful geometric patterns but rather a specifically urban scene.

As I dug further into the urban art of Çatalhöyük, Ian Hodder appeared at the forefront. Hodder's first major and still ongoing archeological excavation, his major move from theory to practice, was in Çatalhöyük, following in the footsteps of his archeological and theoretical mentor, James Mellaart. Mellaart started the excavation of the site in the 1960s and had published a widely read and controversial article (Mellaart 1964) that claimed Çatalhöyük was the first Neolithic city, a "metropolis" originally consisting not of farmers but mainly of hunters and gatherers who were also engaged in long-distance trade.[14] One of the vital commodities for Stone Age peoples was obsidian or volcanic glass, an obvious reference to the red mountain erupting in the wall painting.[15] I would later find out that obsidian was polished at Çatalhöyük to create the oldest known handcrafted mirror, as much a form of self-contemplation as the wall painting itself.

Obsidian and the wall painting also connected my research, to my great surprise, to Jane Jacobs and, through her observations on Çatalhöyük, to her astonishing idea that cities and the urbanization process may have been the primary generative source of creativity, innovation, and societal development for the past 12,000 years. Without cities, Jacobs argued, we would all be poor.[16] We would have remained nomadic hunters and gatherers as we had been for millions of years before settling down in one place. In *The Economy of Cities* (1969), Jacobs built on Mellaart's work, hypothesized an even earlier start to the urbanization process in a place she called New Obsidian, and developed the idea that the 'spark' of urban economic life was the primary cause of all economic growth and change, including the full-scale development of agriculture and animal husbandry, along with many other specialized production activities.[17]

Jacobs's economic spark was her answer to the magisterial anti-urbanism of Lewis Mumford's *The City in History* (1961). It also represented one of the boldest assertions of urban spatial causality in the twentieth century, but like the spatial assertiveness of Lefebvre and Foucault in the same period it would be misunderstood and even more quickly dismissed as fanciful exaggeration. It would, however, be picked up again by such economic geographers as Allen Scott and Michael Storper and resurrected as a Nobel prize-worthy breakthrough in economic theory by a new and growing group of geographical economists who now speak of the stimulus of urban agglomeration as Jane Jacobs externalities.[18] These urbanization economies, as they were earlier called, with their vivid evocations of urban spatial causality, are beginning to enter the economic textbooks as the primary cause of economic development in the world today.[19]

The debates on Çatalhöyük and what I have called "synekism," the stimulus of urban agglomeration (Soja 2000), link the spatialization of archeology to the frontiers of contemporary spatial thinking in urban and regional economics, and to what may be the most revolutionary breakthrough to have emerged from the Spatial Turn: that cities and in particular their social spatialities are today – as they have been for the past 12,000 years – the primary force for artistic creativity, economic innovation, technological change, and societal development. Twenty years ago, such an idea would have been inconceivable, even absurd. Today it is at the forefront of the Spatial Turn.

The Nobel Prize-winning economist Robert Lucas's (1988) bow to Jacobs, implying that she too deserved a Nobel Prize, was largely absorbed aspatially into the emerging conceptualization of human capital. A few years later, a small band of geographical economists and urban and regional economists and geographers began to raise the idea again, but at first it too was largely ignored or deflected in the growing attention given to notions of social capital. Over the past decade, however, as spatial sensibilities spread more widely than they ever have in the past century and a half, a respectful acceptance has begun to consolidate. The next step is likely to be the emergence of a powerful concept of spatial capital, giving urban spatial causality and the socio-spatial dialectic their long-deserved recognition.

Law and spatial justice: a final note

For many years, spatial thinking and geographical perspectives have been particularly influential in law and critical legal studies. Nicholas Blomley, a geographer, led the way early on in *Law, Space, and the Geographies of Power* (1994) and later began a productive interaction with two legal scholars, Richard Thompson Ford and David Delaney, his co-editors of the comprehensive *Legal Geographies Reader* (Delaney *et al.* 2001). Concepts of territory, jurisdictional boundaries, zoning, property rights, public space, and racial geographies are at the core of this collection of papers and the work of its three editors, and have sparked a continued partnership between law and critical spatial studies.

Emanating from this cross-fertilization of law and geography has been another stream of scholarship rethinking in increasingly spatial terms such concepts as justice, democracy, citizenship, governmentality, and human rights. It is here, where my current research and writing is concentrated, that I wish to conclude this essay on the Spatial Turn. I started writing on "Seeking Spatial Justice" in

Postmetropolis, and have been working and lecturing on this subject ever since, building on such sources as John Rawls and his legal theory of justice, the critical expansion and spatialization of the concept of justice found in David Harvey's *Social Justice and the City* (1973), and several more recent extensions by Iris Marion Young (1990 and 2000) and Gerald Frug (1993) on the related notion of democratic regionalism.

The concept of spatial justice is not meant as a substitute for social, economic, or any other form of justice, but represents and encourages a strategic and theoretical emphasis on the specifically (and often neglected) spatial aspects of justice and injustice, including how they are embedded in urban spatial causality. This points to questions of distributional equity but also focuses on making the processes producing unjust results more equitable or, paraphrasing Harvey's notion of territorial justice, seeking just outcomes that are justly arrived at. As Mustafa Dikec (2001) notes in "Justice and the spatial imagination," there is injustice and justice built into the spatiality of our lives and a relevant spatiality to (in)justice itself. This conceptualization of spatial or territorial justice, like the Spatial Turn, started to develop in the 1960s and early 1970s but advanced very little until the mid-1990s, when it began to be revived and reconceptualized within the context of contemporary urban spatial politics (Soja 2007).

Of particular importance and relevance to the theory and practice of spatial justice today, as well as to advancing the Spatial Turn, has been the rediscovery of Lefebvre's idea of *le droit à la ville*, or the right(s) to the city, once the rallying cry for the events of May 1968 in Paris (Lefebvre 1996). In perhaps the strongest and most successful extension of the Spatial Turn into political practice, the notion of rights to the city, concretizing calls for universal human rights by embedding them in specifically urban spatial contexts and causalities, has been mobilizing multi-scalar political movements, reaching from community-based organizations and coalitions struggling for better housing and public transport to regional and national efforts to reduce spatial inequalities in wealth and well-being and achieve more democratic distributions of power, to global "civil society" movements aimed at peace and justice in international and environmental geopolitics (World Social Forum 2005).

More than ever before, these multi-scalar efforts are consciously and strategically spatial struggles based in a new critical spatial imagination that draws on what Lefebvre and Foucault were arguing more than thirty years ago: that 1) we live in socially produced spaces that are predominantly urban and almost entirely urbanized; 2) because they are socially produced rather than naturally given, these urbanized spaces are subject to being changed through social action; 3) the urban geographies in which we live produce powerful negative as well as positive effects on our lives; and 4) the injustices and oppression that are built into our geographies can become a strategic force for mobilizing and organizing innovative forms of spatial praxis aimed explicitly at achieving greater spatial justice and "glocal" democracy, stretching across all the nested geographical scales in which we live. Joining together to foster greater spatial justice may in the end be the best way to promote and expand the Spatial Turn today and in the future.

Notes

1 If you have seen the film *The History Boys*, based on the award-winning play by Alan Bennett, you will remember that the title referred to the eight brightest students at a grammar school in northern England who are assiduously educated to obtain entry to Oxford or Cambridge, which all successfully do. They share a deep interest in history and debate its merits and complexity adeptly, assisted by the history teacher, seemingly the most balanced and perceptive of the teaching staff. But do you remember the primary villain, a sniveling headmaster who basks in the students' accomplishment and seems never to understand their cleverness? While interviewing someone he hopes can help the students spice up their test essays, he admits that he did not go to Oxford or Cambridge. "It was the 50s after all," he said. "It was a more adventurous time. I was a geographer and went to Hull." One could almost hear the audience snigger, a geographer at a minor university being almost as far away from the history boys as one can be. It was the almost subliminal matter-of-factness of the historian's superiority that came through most clearly to me.

2 I returned to my Bronx neighborhood in the early 1970s after being away for more than a decade only to discover that the tenement building in which I grew up had that day begun to be razed. The wrecking ball had just taken one swipe before work ended, scooping out the corner of the fifth floor that was my family's home. Once so solid, my early lived space had melted into air, a prelude to a series of urban regeneration projects that would pave over my street corner, erect and then raze a string of tower blocks, and eventually produce the bizarre suburban row houses that exist there today.

3 I would first visit Andorra in 2007, along with Abel Albet and Nuria Benach, with whom I would share my autobiographical notes to help in their preparation of a book on my work in Spanish (see Benach and Albet 2008).

4 I could run for a bit with my colleagues and their students. I even did a network analysis of connectivity in the Nigerian transport system that identified a zone of peak accessibility that I suggested should be the place for a new capital city. Ironically, and without reference to my work, the Nigerian government would later shift the capital from Lagos to Abuja, right in the center of my peak access zone.

5 I want to emphasize here and elsewhere that the critique of social historicism does not entail an antagonism to history and historiography, nor does it represent the assertion of a privileged spatialism or spatial determinism. The aim is to achieve a critical (re)balancing of space and time, spatial-geographical and temporal-historical perspectives.

6 Many have taken Foucault's heterotopia to mean a specific kind of space. My view is that it represented a way of looking at any space, a particular perspective rather than a particular spatial form. Supporting my view is Daniel Defert, a close friend and associate of Foucault, in his essay accompanying the online advertisement of a CD recording of Foucault's 1966 radio lecture on his notion of heterotopology. To access this essay, see http://www.abeillemusique.com/produit.php?cle+10832.

7 Lefebvre's former student, Manuel Castells, called him the left-wing version of the Chicago School ideology in his redefinition of the "urban question" (Castells 1972).

8 Michael Dear (2000, 2002) has taken on the tasks of defending his brand of postmodernism against all others, and publicizing his own interpretations of Los Angeles and the Los Angeles "school." They are not mine.

9 Economics, after a peripheral flirtation with positivist geography and regional science, has experienced perhaps the most emphatic Spatial Turn in recent years, at least in terms of developing the strongest assertion of the power of urban spatial causality. See the discussion of Jane Jacobs below. Anthropology in its new urban focus and continued interpretations of culture has been creatively spatial as well (Gupta and Ferguson 1997; Ferguson and Gupta 2002). Sociology, especially in the US, has experienced a "rise of spatial metaphors" (Silber 1995) but on the whole has been relatively nonreceptive if not antagonistic to the Spatial Turn, especially in the relative decline of

specifically urban sociology and the deep suspicion of any hint of urban spatial causality, due perhaps to continued reactions to the crude causality of the old Chicago School. Also relatively weak as yet is the spatialization of Psychology and Political Science, but for the latter see the interesting work of Kohn (2003). While written by a geographer-planner, Engin Isin's *Being Political* (2002) is the most insightful assertion of the essential urban spatiality of politics, drawing on Aristotle as well as Foucault and Lefebvre.

10 As I was finishing this essay, I discovered a most remarkable website for surveying the literature on space and place, http://pegasus.cc.icf.edu/~janzb/place. htm, organized and maintained by Bruce Janz of the Department of Philosophy, University of Central Florida, it contains nearly a thousand references in around 40 different fields, ranging from queer theory to animal biology to urban and regional planning. Not all references apply to what I am calling the Spatial Turn, but the listings can keep the spatial thinker and bibliophile occupied for many hours.

11 See the essays on "Men in Space" and "Boys Town" in Deutsche (1996).

12 At the Tate Modern in 2000, I met with the then-curator Iwona Blazwick and we discussed the growing interest in space and urbanism among artists. At that time, I thought these interests primarily involved increasing attention to architecture, but instead the curator was deeply informed of my work and that of Foucault, Lefebvre, David Harvey, and Fredric Jameson. In the conversation that ensued, I mentioned my playful characterization of John Berger as an "art geographer" and was told that the Tate was advertising for two art geographers to enhance their spatial communications with surrounding neighborhoods and the Greater London area as well as to help in organizing an upcoming exhibition on global cities.

13 For more details on the early connections between Archeology and Geography, see Hodder and Orton 1979; Wagstaff 1987; Gamble 1987; and Clarke 1992.

14 Why farmers would cluster in permanent settlements of 2,000 and more inhabitants is not easy to explain. Contemporary archeological evidence suggests that the first urban settlers in Southwest Asia around 12,000 years ago were almost surely predominantly hunters and gatherers who settled down largely to make long-distance trade more efficient, although this interpretation remains controversial among most archeologists and prehistorians.

15 Some doubt has been cast on what exactly is depicted in the mural. Even Hodder leaves open some small possibility that the painting may not be of the settlement itself (see the fascinating Çatalhöyük website and *The Leopard's Tale*, 2006). A recent television documentary on the Neolithic claimed that the volcano, thought to be Hassan Dagh, is too far away from Çatalhöyük to be seen, thus making it possible that the painting actually is of an older and much closer settlement, such as Asikli Höyük. If one climbs to the top of the mound (höyük) at the site, as I did in 2005, nearly all doubts about what is shown disappear. Looking south, one can imagine the long side of the oval ancient settlement in front of you, with a volcanic mountain shaped very much like the one in the painting floating on the horizon. Only the mountain is Kara Dagh not Hassan Dagh, contrary to everything that has been written about the mural. Otherwise, the mural is a direct representation of what would have been seen from the top of the settlement of Çatalhöyük.

16 The full quote, taken from a *Los Angeles Times* interview (Proffitt 1997), is "Cities are the mothers of economic development, not because people are smarter in cities, but because of the conditions of density. There is a concentration of need in cities, and a greater incentive to address problems in ways that haven't been addressed before. This is the essence of economic development. Without it we'd all be poor."

17 Jacob's assertion that "cities came first," that cities were necessary for the production of an agricultural surplus rather than the other way around, remains controversial and is not widely accepted by even the best contemporary archeologists and prehistorians. It is becoming increasingly clear, however, that the first urban settlements had few if

any farmers when they were initially established, that agricultural production primarily expanded to meet the needs of growing urban agglomerations, and that the economic development and wealth of human societies has been more shaped by the urbanization process from around 12,000 years ago to the present than any scholar other than Jacobs and her followers had ever imagined. For further elaboration of these arguments, see Soja (2000, 2003) and Blake (2002).

18 The earliest identification of Jacobs's contributions to economic development theory was by the Nobel Prize-winning economist, John Lucas Jr. (1988). See also H. Harrington, "Review of The Economy of Cities," and D. Nowlan, "Jane Jacobs among the Economists" (Allen 1997).

19 One even more widely popular offshoot of this new geographical economics and the revival of Jane Jacobs is the notion of creative cities, a target that has captured urban planners around the world as a means of taking advantage of the stimulus of urban agglomeration. The key work here has been Florida (2002).

3 Spacing movements

The turn to cartographies and mapping practices in contemporary social movements

Sebastián Cobarrubias and John Pickles

> Something unique is afoot in Europe, in what is still called Europe even if we no longer know very well what or who goes by this name. Indeed, to what concept, to what real individual, to what singular entity should this name be assigned today? Who will draw up its borders?
>
> (Derrida 1992: 5)

> The diagrammatic or abstract machine does not function to *represent*, even something real, but rather constructs a real that is yet to come, a new type of reality.
>
> (Deleuze and Guattari 1987: 142)

The development of global resistance movements has been filled with an expansive spatial imaginary and vocabulary. This has been particularly true since the multiplication and spread of these movements after the first officially declared "Global Days of Action" in 1998 (see People's Global Action in n. 7 below). The multi-issue and transnational nature of this movement, a "movement of movements" (Bergel 2002), reflects its amorphous and flexible nature (Graeber 2004). Central to these movements has been an intense networking and exchange of ideas and practices among movement groups striving to generate an alternative spatial politics of public and private lives (see Ortellado 2002; Marcos 1997; Katsiaficas 2004). Examples of this generative spatial thinking include: issues of "scalar sensitive" politics and the links between the "local," the "global" and everything in between (Prokosch and Raymond 2002); the creation/production of spaces such as the World Social Forum and its unique grassroots popular global diplomacy (World Social Forum 2005); the occupation or reappropriation of spaces through squats, land takeovers, temporary autonomous zones and Reclaim the Streets actions (Notes From Nowhere 2003); and reinvigorated ideas of opening new commons and anti-enclosure struggles (De Angelis 2003; Midnight Notes Collective 2004).

This thinking and analysis often arises not only from efforts to understand global processes and the constitution of corporate power and empire, but also from the "need" to articulate a globalizing identity of struggle among such disparate actors (Solnit 2004; Notes From Nowhere 2003), what Derrida (1994) referred to as the "New International." Thus, on the one hand global resistance movements have been attempting to re-envision the "powers that be" and "what they're up against" through different analyses and metaphors ("the fourth world war", "empire", networked capitalism working through key international institutions and states, etc.), productive metaphors that have generated new ways to understand shared struggles manifest in different ways in different places and through distinct spatial strategies and practices. On the other hand, they have also been attempting to envision new worlds and spaces – "Another World is Possible" or "A World Where Many Worlds Fit" – to reanimate concepts such as transnational/local political spheres and convergence spaces (see Routledge 2003; Featherstone 2004) or alternative or utopian spaces within the space-time of mass demonstrations (Pérez de Lama 2004).[1] In these and wider social movement contexts there is emerging an acknowledgement that, in part, the battle for new worlds is a battle over space and the production of spatial imaginaries.[2]

Understanding how spaces and spatial imaginaries affect movements as well as how movements create spaces of different sorts can help us understand more complex movement dynamics and unevenness in movement development (Lefebvre 1991). Such creative effects of movement action may involve: ephemeral moments of excess at a mass demonstration that can spread like a meme across locales (similar to the eros effect in Katsiaficas 1987); abeyance structures that help provide continuity in moments of demobilization (Taylor 1989); movement infrastructures that provide communications or survival mechanisms to participants (alternative media, trade networks, etc.); and/or institutions that may then articulate with the state and negotiate their own space in governance structures.

Social movement studies as a subfield has recently returned to questions of space and spatial practice, perhaps most notably through work on geographies of resistance (Pile and Keith 1997), global justice movements (Routledge 2003; Featherstone 2003), and recent work on anti-nuclear movements (Miller 2000).[3] In the 2000 Mobilization encounter focusing on the role of space in contentious politics and the practices of social movements, Tilly (1998), Martin and Miller (2003), Marston (2003), and Wolford (2003) each drew on aspects of these traditions to open a conversation between the more taxonomic sociological approach to movements represented there by Tilly and the more dialectical and relational analyses of the socio-spatial representation by the geographers. Each, and in conversation, has, as a result, expanded the lexicon and conceptual apparatus for thinking geographically about the social in social movements. Each points in slightly different directions and each works differently with concepts of space, the social relations of spatiality, socio-spatial dialectics, and spatial practices, but all suggest some important ways in which a post-vanguardist politics attentive to contingency and local specificity must reshape thinking about social movements around different configurations of space and place (see Escobar 2001 for a broader reading of the significance of contingency and local specificity).

For our present purposes we have focused on a more limited context in which theories and technologies of space and spatial mapping are generating particular interest in and among the social movement activists themselves. We are interested not only in what it means to think cartographically about social movements, but also what it means when social movements themselves think and act spatially and more specifically cartographically. To do this we have engaged with several movement collectives that have explicitly adopted the practices of cartography and map-making as one way to further pursue their political activities. Such critically engaged cartographies have begun to multiply in recent years in social movements and art-activist groups. Some of these are cartoon-like maps, others are street protest maps which designate targets, safe zones, or map out areas for differing levels of physical militancy (Figure 3.1). But a newer wave of these activist maps that are more theoretically and analytically engaged has also emerged and seems to be spreading.[4] Here we focus on two groups (Bureau d'Études/Université Tangente and Hackitectura), but the number of such groups is growing rapidly, including parallel groups in Barcelona (De Que va Realment el Forum) and Rome (Transform!) which map the multiple types and sites of conflict in the city; projects such as "They Rule" (www.theyrule.net), which uses the internet to create network maps of corporations, their executive boards and political administrations in the USA; and Migmap, which maps the complex networks and interrelations among actors, discourses, institutions, and places and practices governing migration and Europeanization (http://www.transitmigration.org/migmap/).

Cartography may seem an unusual beginning for thinking about the reconfiguration of social, economic, and geopolitical identities. Historically, cartography has been associated with the imperial projects of the last several centuries by mapping *terrae incognitae* in order to facilitate material and cognitive conquest. Cartography may even seem an irrelevant subject. While it is argued that more maps are being produced now than at any time in the known past, it seems that the majority of these are classically Cartesian and dedicated towards mapping "objects" such as glaciers, streets, military targets, and potential markets.

The relevance of cartography to projects of emancipatory politics, social theory or critical political economy can seem to be limited at best. After all, so much of the new mapping practices and technologies actually seem to only deepen the power of existing institutions, or at least depend on the kinds of investments these institutions (especially the military) make in new visualization capacities and approaches (Virtual Reality, 3-D, high-resolution space mapping, etc).[5] Even in participatory mapping and GIS practice, one may rightfully be skeptical of the emancipatory potential of participatory mapping in which, as Denis Wood (2005) argued, the deployment of terms like "public" "participation" "geographic" "information" "systems" is not about public practices but about the construction of new technical intermediaries and consultational instruments, not about the participation of citizens in the governance of their lives but about a representational politics that domesticates any threat of participation, not about the "geographic" as a complex nexus of meanings and practices but about the rendering of a landscape of instrumental rationalities, and not about information but about the production and regulation of data.

In such a world, the only apparent relevance for a critical project of cartography might appear to be the discursive "deconstruction" of these maps and the worlds they produce. But, as Fredric Jameson (1991) pointed out in his efforts to

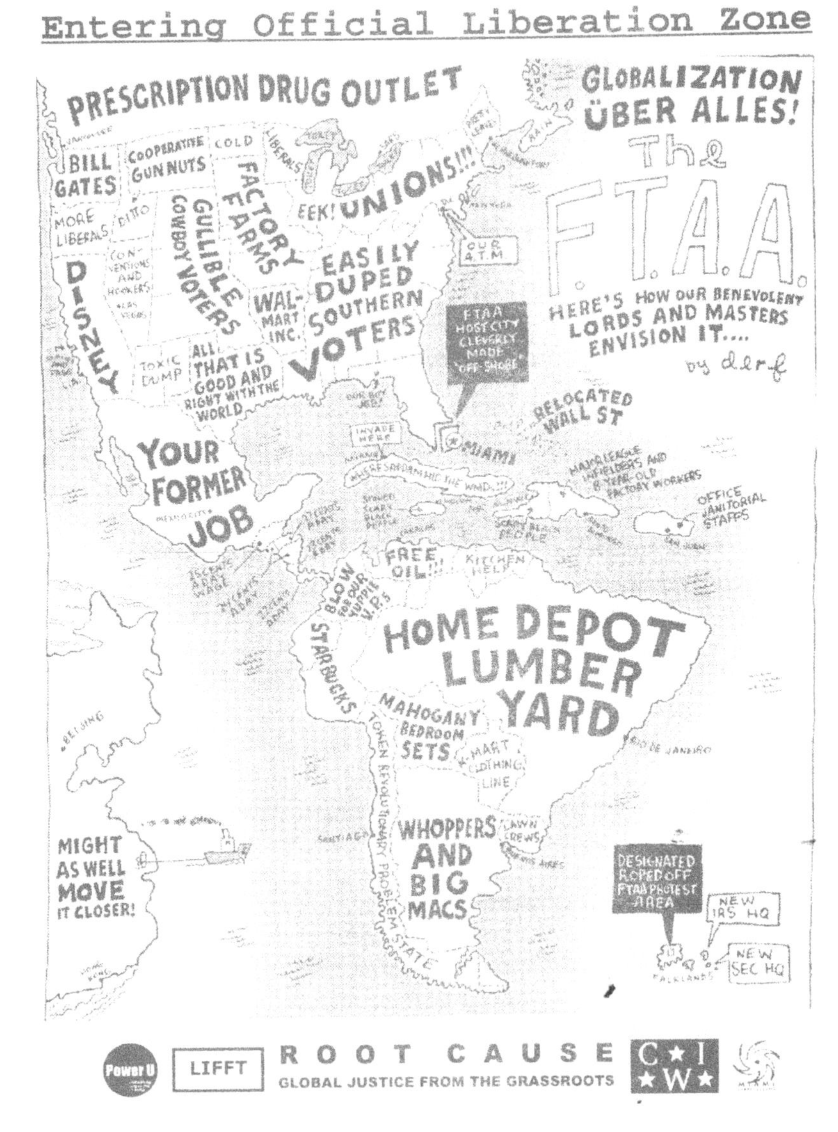

Figure 3.1 Entering the official liberation zone (root cause).

come to grips with an appropriate politics that responds to the dislocationary effects of what he referred to as the "postmodern condition", the unity of a mid-twentieth-century Eurocentric order is broken up into a radical diversity that at the same time masks power and oppression through complexity and disjuncture. Individuals are increasingly unable to perceive the criss-crossing of (now global) power relations that intersect with their daily life. The globalization of capital, trade, and the culture industries project a plethora of "identities" of a society fully imbued with participatory possibilities and choice, but lacking the kinds of "cognitive maps" so necessary for effective political action and the reflective life. This carto-politics seeks to contribute to a political project of global cognitive mapping in order to understand the current spatial logics of capitalism and provide individual subjects with the tools necessary to create alternative trajectories to navigate this system (Jameson 1991: 54).

Such cognitive maps are, of course, always being produced either by design or by default, with some social actors staking out particularly clear positions on their value. The new mappings that most interest us in this work are those Gilles Deleuze – in referring to Michel Foucault – called the "new cartography". The new cartography was to be a mode of spatial thinking that sought not to trace out representations of the real, but to construct mappings that refigure relations in ways that render alternative epistemologies and very different ways of world-making. From this perspective,

> A distinction must be made between two types of science [royal and nomadic], or scientific procedures: one consists in "reproducing," the other in "following." The first involves reproduction, iteration and reiteration; the other, involving itineration, is the sum of the itinerant, ambulant sciences . . . Reproduction implies the permanence of a fixed point of view that is external to what is reproduced: watching the flow from the bank. But following is something different form the ideal of reproduction. Not better, just different. One is obliged to follow when one is in search of the "singularities" of a matter, or rather of a material, and not out to discover a form . . . when one engages in a continuous variation of variables, instead of extracting constants from them . . . And the meaning of the Earth completely changes: with the legal [royal-reproducing] model, one is constantly reterritorializing around a point of view, on a domain, according too a set of constant relations; but with the ambulant [nomadic-following] model, the process of deterritorialization constitutes and extends the territory itself.
>
> (Deleuze and Guattari 1987: 372)

This type of epistemological shift has become the basis for intense analytical and political experimentation and mobilization among these social movements, especially those connected with the global justice movement. The new cartographers are the social movements who – across a very wide spectrum of groups and locations – are rapidly expanding the scope of their spatial practices and their production of new mappings to render new images (and practices) and to render visible their geographies in new ways. But this is not only a representational politics of making the invisible visible (Kanarinka 2005). If traditional cartographies sought to represent the real, new mapping practices seek instead to unmask a new type of real. They produce it, either by rendering it visible as a form of socio-spatial practice and collective action, or by producing alternative imagined (even utopian) spaces to those being built by the state and other transnational actors.

Here the traditional logics of cartography (the logics of tracing and reproduction) are to be transformed by mappings that produce new conceptions of place and region, proliferate the kinds of relations among them, open up spaces for action, and constitute new subjects:

> What distinguishes the map from the tracing is that it is entirely oriented toward and experimentation in contact with the real. The map does not reproduce an unconscious closed in upon itself; it constructs the unconscious. It fosters connection between fields. . . . The map is open and connectable in all of its dimensions; it is detachable, reversible, susceptible to constant modification. It can be torn, reversed, adapted to any kind of mounting, reworked by an individual, group, or social formation.
>
> (Deleuze and Guattari 1987: 12)

The participants in these movement mapping projects are diverse in their backgrounds and career positions, comprising the unemployed, academics, flexible production producers, artists, designers, community and housing specialists, chain store (retail) workers, intellectual labor, among others. In some sense, these groups constitute new research centers of artists, activists, and self-styled hackers working on more productive cartographies directed to a new type of bio-geo-politics. What characterizes these groups, as well as the wider movement networks of autonomous activists of which they form a part, is their focus on the intensification of processes of de-territorialization and reterritorialization occurring in and across Europe, the Americas, and elsewhere. In each of these settings, the role of a spatial politics of enclosure, whether by privatization, dispossession, or financial power, seems to be leading to an increase in the scale and effectiveness of spaces of exclusion and the hardening of territorial-politico boundaries at all scales. Among increasingly dislocated and flexibilized populations, there has emerged a desire to produce new cognitive maps that are crucial instruments of reterritorialization; new machinic forms produced in the act of contesting public spaces and new assemblages aimed at opening alternative possibilities for citizenship.

Mapping is, of course, only one practice among many currently circulating within the intensive networks of European social movements. Many social movements do not engage in mapping or purposely eschew its use. For these groups, if they think of mapping practices at all, it is likely to be in terms of cartography and geographical information systems – the technics of spatial politics – as instruments of surveillance and control. Not surprisingly, some groups worry about the consequences of adopting techniques that historically have been so closely aligned with institutions of power. However, in turning to cartography and mapping, these groups are claiming neither a privileged role for mapping and cartography as technical practices of social action nor are they suggesting that these mapping practices constitute the possibility for any new kind of universal god's-eye logic or epistemology for collective action (Pickles 2004). Indeed, the groups we deal with here explicitly articulate their understanding of mapping against the cartographies, geopolitics, and geo-economies of state territoriality and extra-territoriality.

In referring to the work of Foucault as the "new cartographer," Deleuze (1988: 23–44) pointed to a mode of investigation and writing that sought not to trace out representations of the real, but to construct mappings that refigure relations in ways that render alternative worlds. This new cartography is used by the social movements we discuss to refigure the relations of power that structure socio-spatial life and to remap the social spaces of everyday life in ways that produce new political subjects. The tension between existing and possible spatial imaginaries is thus articulated by Euromovements (a European action research network) as a complex double movement of "mapping the social movements vs. mapping the biopower," and it is this double movement we wish to explore in the sections that follow. In particular, we seek to illustrate some of the ways that these social movements, anarchist collectives, and/or art groups are attempting to think and act spatially in ways that aim at different kinds of mobilization and transformative politics. In their own words:

> Techno-political tools and social transformation constitute an interesting interaction domain. Some of us are calling for the production of "anti-technologies of resistance", arguing for methods that can't be replicated elsewhere, so evanescent that they can't be produced with a guideline to understand it. The exercise of mapping is always under the danger of capturing something [that] wants to be free. How [do] groups producing maps, analyzing and/or producing techno-political tools to systematize large amounts of data deal with those questions? Which are the ethical and philosophical challenges and limits about mapping social transformation, social movements and networks activities? How to reverse control surveillance engineering, how to challenge the pan-capitalism, through mechanisms, as Brian Holmes call them, "grass roots top down surveillance systems"? What is the role of technology and software programation inside the development of this activist research field? We do also get the sensation that [the] maps/visualization field of action is highly competitive, and is fulfilled with private actors and enterprises, how do actors and groups from social movements and hacklabs challenge this?
>
> (http://www.euromovements.info/english/news3.htm)

Mapping has taken on an increasingly important role in this kind of global resistance politics. The number of groups, efforts, and projects that use maps as a technique to engage questions of the global economy, new transnational identities, and changing urbanisms has become dizzying. The styles of maps, the particular goals and the types of cartography on which they draw are as diverse as the groups themselves. As mentioned earlier, some of these maps take the form of community asset mapping, where different participants in a mapping session focus on resources, sites of memory, sites of access, and other capacities on a basic frame of a map with often little more than an outline of a city or region. The maps are usually oriented toward the goals of a particular group or campaign, allowing participants to literally draw links and clarify the connections between disparate resources. Others take the form of the basic tourist or street map, often prepared for a large mobilization, such as the World Social Forums or the Republican National Convention protest of 2004 in New York City. In these cases, the map provides information to participants about targets to be protested, areas that are likely to be sealed by police, the location of safe areas, as well as resources such as information centers, independent bookstores, free wireless access spots, and available medical services. While such maps are designed and distributed for a specific event over a short period of time, they also become something that can be used long after the mobilization. Yet other maps juxtapose seemingly disparate and discrete entities and issues in order to create or make visible relationships and new conceptions of space, the maps themselves becoming a form of community-making through a collective process of production imbued with activist politics. We focus on two art-activist groups that have been actively engaged in these two forms of carto-politics: Bureau d'Études and Hackitectura, respectively based in Paris and Seville. Each is a key referent and inspiration for this growing trend of activist mapping and a resurgence of spatial practice in Europe.

Their "new cartography" is productive, aimed at mobilizing alternative geographical imaginations, expanded spatial practices, and new worlds. Not all movements that use mapping are "new cartographers" in this vein and some may not be explicitly aware or critical of the ways in which cartography and mapping have functioned as forms of state or "royal" science in the past. However, the number and scope of these varieties of cartography and hacking groups is growing and their work is beginning to circulate to great effect. These are groups with antecedents in situationist, schizo-analytic, and a variety of other avant-garde activist and revolutionary practices. They are not social movements based on modernist logics of representation and equivalence, but instead are part of a broader movement of autonomous politics and a kind of nomad thought that "replaces the closed equation of representation, x = x = not y (I = I = not you) with an open equation: . . . + y + z + a + . . . (. . . + arm + brick + window = . . .) . . . Nomad space is 'smooth,' or open-ended. One can rise up at any point and move to any other" (Massumi 2002: p. xiii).

This perspective has consequences for the logics of collective action and political mobilization. The groups attempt to create distinct political infrastructures that can exist parallel to, and in antagonism with, those of the state and political party; by subtracting from the "state", the space of the EU, or of capital, they proliferate spaces in complex and dynamic interaction with each other and in ways that highlight their contingency. Theirs is a struggle for a politics of what Deleuze and Guattari called the rhizome that "unlike trees or their roots . . . connects any point to any other point, and its traits are not necessarily linked to traits of the same nature; it brings into play very different regimes of signs, and even non-sign states" (Deleuze and Guattari 1987: 21).

Bureau d'Études and the mapping of neoliberal Europe

In 1998, with the exhibition Archives du Capitalisme, Bureau d'Etudes started producing organizational charts showing the proprietary relations

> between financial funds, government agencies, banks and industrial firms. A number of these graphic charts, or "organigrams," were deployed as part of an installation including black-and-white photographs of heads propped up on wooden pickets (presumably CEOs), as well as a scale model of a proposed new parliament building, to articulate the voting rights of those with real power in today's society. The exhibition was an autonomous project in an artist-run space, at the time called the "Faubourg," in the city of Strasbourg. For a subsequent show entitled Le Capital, mounted by Nicolas Bourriaud in the city of Sete, an organigram detailing the relations between the French state and a panoply of major transnational corporations was blown up to wall size. Squares and rectangles of varying proportions, each identified with a name (Societe Generale, Dresdner Bank, Mitsubishi, Pirelli, etc.) were connected with a labyrinth of elaborately traced channels, printed in black against a white ground. The result was something like an all-over painting for the computer and finance-obsessed 1990s: an aesthetics of information.
>
> (Holmes 2003)

Bureau d'Études mapping projects began around 1998 with a collection of political art called the "archives of capitalism". The archives emerged out of their engagement with unemployed and squatters movements and included maps and flowcharts of economic networks and powerful individuals in the Strasbourg region produced as a form of public/political art. After several other collective projects in France, the group began to question how to break out of the typical gallery-museum art circuit. In coordination with other artists, they founded the "Syndicat Potentiel" (the "Potential Union") to engage with issues of unemployment, casualized labor, and culture workers. Both groups (Bureau d'Études and Syndicat Potentiel) are tightly networked with other activist groups and have strong connections with squatter movements throughout Western Europe (Holmes 2003: 1–2). One of the central goals of these groups is the production of "autonomous counter knowledges" and – drawing on the work of Marcel Mauss on the gift economy and the rediscovery of Mauss in France after the general strikes of 1995 – they have been particularly active in trying to frame discourses and practices of an economy of the "free" (free goods, services, etc. see Manifesto of the Université Tangente, as well as Holmes 2003; Graeber 2002a; http://www.revuedumauss.com/).[6]

Reflections on the changing nature of the economy and the need to better understand the new forms of international corporatist and state solidarity led to a long-term engagement with cartographic representation. Particularly after the first "global days of action" in 1998[7] and the widespread emergence and acknowledgement of global resistance movements, these cartographic experiments expanded and resulted in an expansion of the Bureau d'Études out of the gallery-museum dynamic and into more open circulation (Holmes 2003: 3). The maps show dense networks of institutional actors in regional and global economies, and they are intended to function as shock tools to incite conversation and analysis.[8] They have been distributed at events such as the European Social

Forum, counter summits against important international institutions, No Border camps against migration policing, and other movement spaces (see http://utangente.free.fr/). The group has produced over a dozen major maps, as well as accompanying texts, and these have now been used in a wide variety of settings and by different groups in global resistance and anti-capitalist movements.[9] The mapping work of the Bureau d'Études has inspired individuals and collectives in different countries (especially in Spain, France, Germany, Serbia, the UK, the USA, and Canada) to experiment with similar forms of map-making as intervention (see Université Tangnete 2003; Kuda 2004).

> [T]hese maps present an excess of information, shattering subjective certainties and demanding reflection, demanding a new gaze on the world that we really live in. These are synoptic visions of the contemporary, transnational version of state capitalism, as constructed "by collusion between specific individuals, transnational corporations, governments, interstate agencies and 'civil society' groups."
>
> (Holmes 2003)

They make visible the institutional patterns that have structured themselves in an overarching, terrifyingly abstract space, almost totally beyond the grasp of the democratic counter-powers formerly exercised within the purview of the national states, and indeed, almost totally invisible – at least until recently, when the communicative possibilities have allowed a certain measure of "cognitive mapping" to be performed by inhabitants (Holmes 2003).

Bureau d'Études maps in the "European Norms of World Production" series are particularly instructive in showing how it is trying to use a form of mapping to challenge accepted categories of Europe.

> Absent from the local landscape, invisible to the naked eye, a labyrinth of laws and standards lends tangible form to our existence. The European Union . . . is an attempt to produce the world we live in. The instruments it uses are norms: industrial standards, territorial models, ideological guidelines, truth criteria. These become the second nature of an expanding, accelerating drive to make this vast, unpredictable human region into a playground for capitalism. Sophisticated services have now arisen to lead corporations through the tangle of agencies that their own lobbies helped to create, as a smokescreen to hide and further their own interests.
>
> ("European Norms of World Production")

The map is polemical, representing a dizzying array of institutions, actors, personalities, organizations, and movements. It comprises three parts. On one side of the map is the "Normopathic Complex (Europe)" with a series of EU institutions, nation-state institutions, corporations, lobbies, think-tanks, personalities, policy initiatives, regulatory agencies, court systems, police forces, and a wide array of norms and laws that facilitate its expansion (Figure 3.2). Links are drawn between the different elements to create a sort of network map

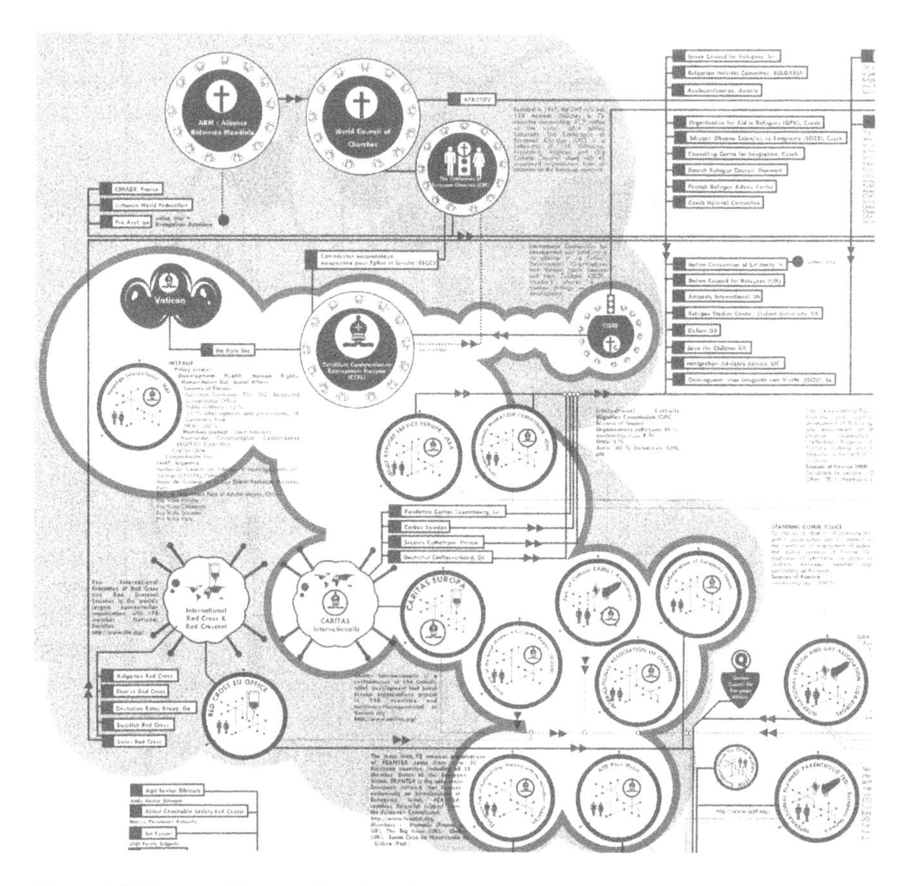

Figure 3.2 Normopathic complex. http://ut.yt.t0.or.at/site/carte/normopathic_complex2002 A1.pdf

of corporate, state, and regulatory power. Part of the second side/layer of the map focuses on "organized civil society" and includes NGOs, EU committees on civil society, and non-industry policy platforms that are linked in multiple and complex ways with the nation-state, the EU, and industry (particularly indirectly through secondary groups, such as task forces, think-tanks, etc.) (Figure 3.3). The final layer of the map is "Inklings of Autonomy" and includes a wide array of social movement activity. The movements are purposefully represented with porous boundaries so that their concerns and commitments bleed from one into another: "anti-prison", "abolition of the state", "re-appropriation of public goods and services", "heterodox research centers," etc. (Figure 3.4).

For the Bureau d'Études, the new Europe is being produced precisely through the interplay of these three layered and complex assemblages of institutions and

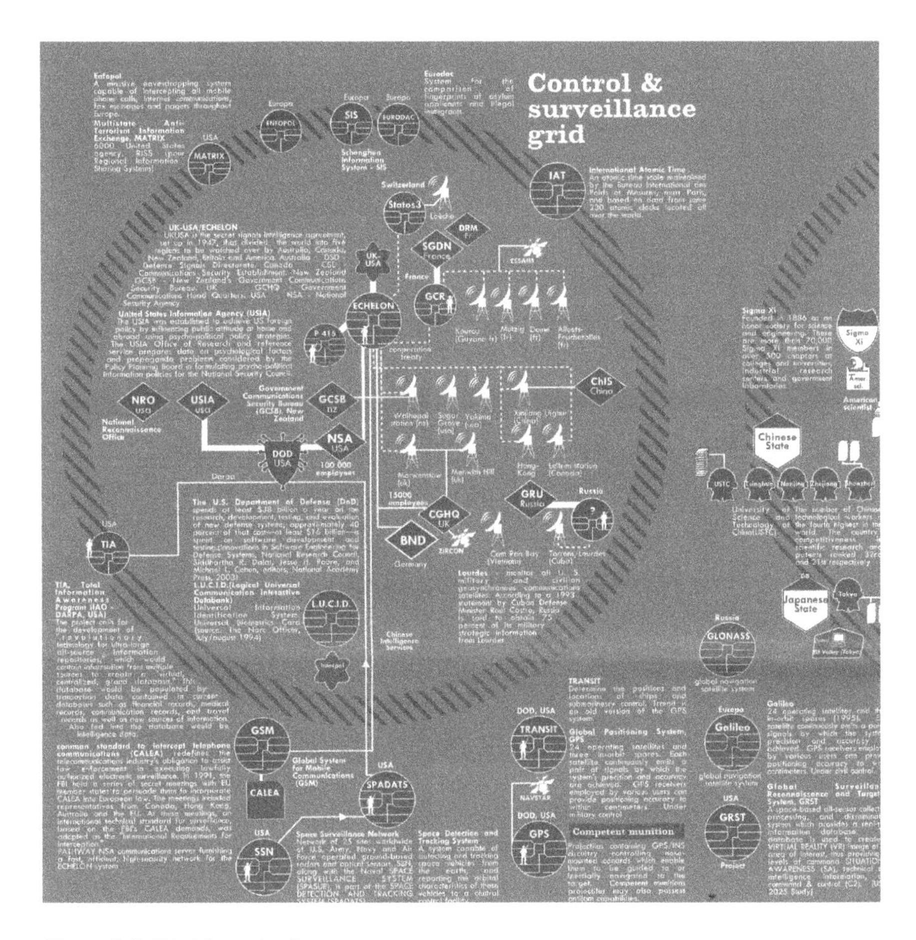

Figure 3.3 World monitoring.

actors. Bureau d'Études thus deploys a cartography of assemblage, beginning from one location (law, institution, or corporation) and piecing together connections and networks in ways that relocate those same laws, institutions, and corporations in a larger framework. In this way, the map locates and materializes the abstractions "EU" and "Europe" by embodying them in specific and networked institutions and laws. Bureau d'Études refers to these as "poles in the reorganization of the terrestrial production line" (Table 3.1).

The resulting maps are both concrete tools for developing specific tactics *and* a form of ideology critique that destabilizes the seemingly solid and autonomous institutions. The institutions that depend so fully for their "solidity" on their existence in the web of relations that produces a particular "Europe" are denaturalized. Instead of being seen as a hierarchically organized set of power/knowledge dispositifs, this new Europe is now rendered as a complex articulation and mesh

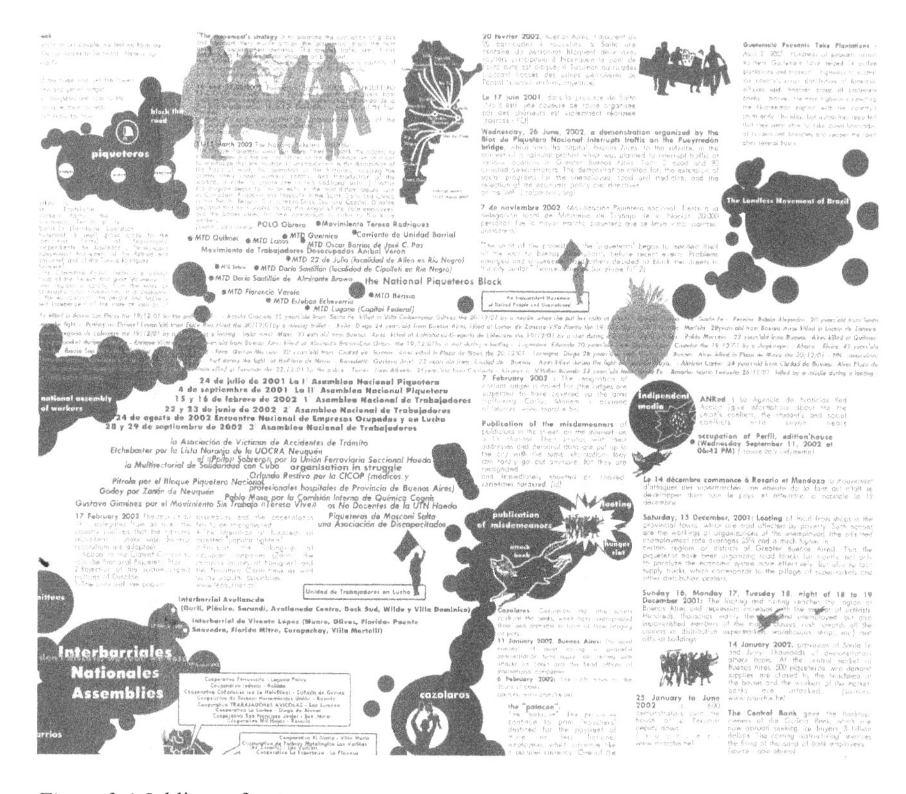

Figure 3.4 Inklings of autonomy.

of contingent actors and institutions. Indeed, in "Flowmaps, the Imaginaries of Global Integration," Holmes (2004b: 5) suggests that Bureau d'Études maps should be seen as attempts to visualize spaces of flows where specific actors such as states, corporations, investment funds, armies, think-tanks, lobbies, powerful family lineages, media groups, transnational religious organizations, and social movements are mapped as nodes through which elements such as knowledge, money, and media flow, not in a unidirectional nor stable manner but with stronger or weaker connections.[10]

European Norms of World Production has primarily been used as a workshop tool for teach-ins. Typically it has the immediate effect of impressing people with its dizzying complexity and generating in its users a sense of awe for its visual complexity and institutional reach. But it is then used to stimulate discussion about how to use the map and what can be done with it; to understand how concrete and contingent are the institutions of the new Europe, to identify the many points at which this new "Europe" is vulnerable and at how many levels it is literally pieced together. It has certainly produced a more directed dialogue about

Table 3.1 Bureau d'Études war times chronicles: poles in the reorganization of the terrestrial production line

Pole 1 – technological and military pole: telecommunications and armaments firms, secret services, states, think tanks, networked police forces, sects, cable networks, satellites, universities.
Pole 2 – pole of control over world financial flows: banks, private and public pension funds, transnational firms, elite leadership bodies, investment companies, lobbies, states, mafia networks, off-shore financial zones, international public organizations.
Pole 3 – pole of control over communication and information, or more generally, over the industrial production of decoys: transnational media firms, ministries of culture and communication, newspaper and magazine, television, radio, publishing and consumer-goods companies, advertising firms, fashion and clothing firms, supermarkets, stores, malls, cinema conglomerates, sects, secret societies, author's groups, counseling and expertise services.
Pole 4 – pole of control over the production and productive optimization of life: transnational firms, universities, start-ups and biotech companies, copyright and patent firms.
Pole 5 – pole of control over access to the planet's natural resources, in particular water and oil: states, major oil companies, transnational firms, lobbies, international organizations, mafia networks, Françafrique network, cartels, pipelines.

Source: http://utangente.free.fr/anewpages/ac.html

what the EU is and how to challenge it, and it has also helped spread the idea of mapping as a tool to map out power and collective struggles. Bureau d'Études is now working with other collectives to develop online map generators that can enable more participatory mapping practices and which will allow input from different groups and campaigns.[11]

Hackitectura and deleting the borders of fortress Europe

The activist mapping efforts on which we are focusing are, at one and the same time, attempting to navigate the changing contours of something called "Europe" within the context of global resistance movements and also attempting to create new geographies and spatialities that prefigure the sorts of social relations these movements would like to enact, often taking advantage of the very same reconfigurations of European "space" that are afoot (such as the reworking of borders) as a way to inject and pursue their own projects. Hackitectura has been particularly important in this respect.

Hackitectura is based principally in Seville, Spain, though it works closely with a network established throughout the region of Andalusia and parts of northern Morocco. It emerged formally in 2004 at the intersection of several preexisting projects (mobilizations, maps, texts, websites) and has been extremely active since then. As the name implies, it is a network of hackers, artists, and architects and others involved in activism in the area (see www.hackitectura.net). Hackitectura is situated in some very interesting and complex nodes of social

Figure 3.5 Mapa de la Sevilla Global / wewearbuildings *et al.* 2002; http://www.hackitectura.net/osfavelados/sevilla_global/mapa_highrez.html

movement activity in the area, including independent media and technology initiatives, migrant rights organizations, and emerging efforts at organizing around flexible labor markets.[12] Their early mapping projects resembled street maps, although their goal was to chart the diversity of protest actions and events at one of the large anti-globalization protests. Individual groups could plot their different actions on a larger master copy and it could be used as an infopoint at a protest convergence center or reprinted for handheld use. From these efforts at showing where protests were happening, later efforts focused on how to translate discussions of the effects of "globalization" on a city-wide scale. In Seville, on the eve of a large EU summit and concomitant protests in 2002, a collaborative mapping project was carried out with different community organizations to try to represent different effects of globalization on the city (Figure 3.5).[13] The project drew on Zapatista frameworks of understanding, particularly those from the communiqués on "Seven Loose Pieces of the Global Jigsaw Puzzle" (Marcos 1997) and the "Fourth World War" (Marcos 2001). In this case, the map is part of a broader collective project of revisioning the city, a project that lasted much longer than the action itself.

The success of these efforts led to projects that attempted to challenge historical and official spatial conceptions of the EU–Africa or North–South border between Spain and Morocco. As struggles for immigrant rights and against

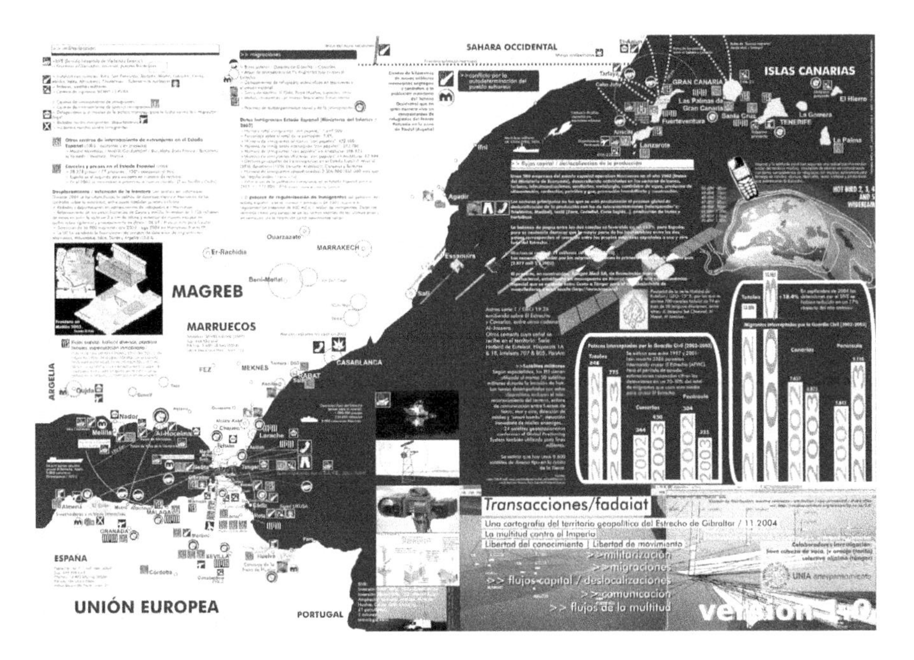

Figure 3.6 Cartographies of the Straits of Gibraltar. http://mcs.hackitectura.net/show_image.php?id=593

precarious/casualized work grew across the country, one flashpoint for their articulation was this border and this, in turn, made it more urgent that alternative spatial logics and imaginations be mobilized. The creation and use of alternative cartographies of the border became one strategy for beginning to reconceptualize the terrain of the "border regime." In order to carry out these remappings Hackitectura participated in a network between social activists in southern Spain and northern Morocco and together they attempted to reconceptualize the spaces of the border and imagine new ways in which a relational cartography of flows and networks might rework the geographies of alterity and difference, focusing instead on the rich and diverse ways in which the border region constituted a dense regional meshwork. Instead of focusing on maps that delimit Europe, Hackitectura began to refigure the border as a spatially interwoven set of practices, institutions, and technics; the exchange of bodies and goods, transnational and transborder telecommunication and broadcasting spaces, integrated surveillance structures, shipping lanes, and atmospheric and oceanic exchanges of energy and material. These networks mapped out new conceptions of economic change and global flows, as well as innovative ways of thinking about community action at the border (Figure 3.6).

Instead of accepting the border as a fixed entity that separated an "us" from a "them," constraining bodies and movement, the groups involved are mapping the complex networks of flows that make up this "border" region: including flows of capital, police, and detention centers ("geographies of empire") as well as networking between movements, migrants, and new technologies ("geographies of the multitude"). To these they have added new interaction spaces, such as those created by communication technologies that span and connect the region of the straits and facilitate ever tighter networks of contact and coordination on both sides. The result is a map that does not reproduce the border as a space of separation but follows the flows across the Mediterranean in order to articulate the border as a space that is created, inhabited and traversed. In this rendering, the "solid Mediterranean" is no longer understood as a natural boundary, but a lived space of historical continuities and connections (Pérez de Lama 2004).

Spacing social movements and the new cartographers

Alan Watson, European chair of the world's largest public relations firm, Burson-Marsteller has little doubt about the importance to his corporate clients of producing such cognitive "way-finding" maps:

> First of all they've got to start by knowing how the system works. . . . We can advise them on how they should put their argument on paper. We can advise them on which people in the committee would be interested. . . . So that way you build a map for them, a sort of road map of where they need to go, who they need to talk to and what they need to know. . . . We don't do the lobbying . . . What we do is to give the company the information so that they can go and make the case for themselves.

Groups like Bureau d'Études and Hackitectura have long recognized that the need for and interest in such road maps is not confined to corporations. Social movements, art activists, and autonomous groups can all deploy cartography and experimental forms of spatial practice as tools for new forms of geo-bio-politics (see Pickles 2004, 2006).

One influential figure involved in the new radical networks of mapping and mappers is Brian Holmes, who in a wide range of published papers, projects, and actions has mobilized much interest around the possibilities cartography provides for analyses of and responses to the global economy and emancipatory politics. Holmes (2004a, 2004b) seeks to deepen Jameson's global cognitive mapping project through the poststructuralism of Tally (1996) and Bartolovich (1996) and the new materialism of Gilles Deleuze, Felix Guattari, and Michel Foucault (Holmes 2004b).

For Holmes (2004a: 4) the new cartographies work in the spaces between dominant and dissenting maps: "every successful cartography ultimately helps create the world it purports to represent," and it does so in part through the organization of information and the rendering visible of the global economy in new ways (Holmes 2004b: 2). Holmes stresses the "need [for] radically inventive maps exactly like we need radical political movements: to go beyond received ideas and orders, in fact, to go beyond representation, to rediscover and share the space-creating potentials of a revolutionary imagination" (Holmes 2004a: 1).

Holmes compares two styles of maps. The first style is that of a "geographical representation of networked power – a determinate network map which attempts to identify and measure the forces in play" (Holmes 2004b: 7). This map has fixed borders and actors with fairly clear dynamics underlying their relationships.[14] The second form of map is that of "an undetermined network diagram, which opens up a field of possibility or of potential strategy" (Holmes 2004b: 7). Holmes invokes the notion of "diagrams of power" from Deleuze's work on Foucault: "a cartography coextensive with the whole social field." The map does not designate a "static grid" fixed in spaces but rather a productive matrix that interacts across a myriad of "points-human beings" and spaces. This productive matrix coexists alongside and in tension with others operating throughout the realm of the "social". "Deleuze describes the diagram of power as "highly unstable or fluid . . . constituting hundreds of points of emergence or creativity." The aim [of mapping] is to indicate the openness, the possibility for intervention that inheres to every power relation" (Holmes 2004b: 8). Mapping can become a way of visualizing this "meshwork" (see De Landa 2005 and Escobar 2004 on meshwork and social movements). It is in this second form of cartography where Holmes sees incredible potential for new radical mapping efforts such as Bureau d'Études and Hackitectura.

Saul Albert (2003) similarly attempts to push past some of the underlying assumption of traditional "systematic cartography" in his turn to poststructuralism and Deleuzian-like "diagrams of power." In this context, Albert is more interested in linking these possibilities to the actor-network theory of Bruno Latour and John Serres. For Albert, there is in Latour's thinking a similarity between these new cartographies and actor-network theory at the level of their practices and their critique of "representational norms" (Albert and Latour in Albert 2003: 4–6). Albert goes so far as to claim that Latour's example unintentionally chalks an outline around the missing half of Harley's critique: that cartography is potentially an ontological investigation. If it removes its *a priori* assumptions, it becomes a kind of spatial ontology, one that is well equipped with both the tools and methods of constructivist research, and the deontological moral standard of "irreducibility" (Albert 2004: 6).

This deontologized cartography can now function much like John Serres's "quasi-objects" – nodal points of articulation of particular interconnected, but contingent roles, subject positions, and sets of social relations. This is how Serres would view the map, as a formalization of human relations, a representation with which each actant becomes a subject. This is the use of the map as a communicative tool; as successive actants engage with the map, each locates their subjectivity in its representational schema, the "I" is shifted from person to person, between person and multitude, or from multitude to multitude. "Analytical cartography", and the power relations Harley identifies in it, is an example of the "deterministic practices" this use of the map may give rise to (Albert 2003: 9).

"[W]e can approach [these] map[s] of global flows as diagram[s] of power in the Deleuzian sense . . . not simply as 'static grids' but rather 'productive matrices' criss-crossed with tensions. The networks visualized are indeterminate, open to 'a field of possibility or of potential strategy'" (Holmes 2004b: 7). In this sense, Bureau d'Études maps can be seen as "cognitive tools" (Holmes 2003) "responding" to Jameson's call for a global cognitive mapping of the scales and structures of a global system ungraspable to any (according to Jameson) individual subject wandering through it. At the same time, these mappings not only allow for navigation but for Bureau d'Études and Hackitectura, they allow for reappropriation:

> Autonomous knowledge can be constituted through the analysis of the way that complex machines function. . . . The deconstruction of complex machines and their "decolonized" reconstruction can be carried out on all kinds of objects, . . . In the same way as you deconstruct a program, you can also deconstruct the internal functioning of a government or an administration, a firm or an industrial or financial group. On the basis of such a deconstruction, involving a precise identification of the operating principles of a given administration, or the links or networks between administrations, lobbies, businesses, etc., you can define modes of action or intervention.
>
> (Bureau d'Études/Universite Tangente 2002: 3)

Conclusions

Through mapping, Bureau d'Études, Hackitectura, and Brian Holmes attempt to show some of the ways in which grids of power and determination are undermined by their own overdetermined structures and complexities. Each has tried to see in complex grids of power self-organizing lines of flight and real possibilities for alternative social and economic organizations. In so doing, each has also elaborated the implications of Deleuzian materialism for the practices of social movements. Together, they see in traditional fields like cartography new epistemological and political possibilities for producing new subjects and relationships, thus destabilizing hegemony by the material practices of new cultural production. They are not myopic about the power of maps. In fact, Holmes (2002) ends a recent essay with a note of caution, and perhaps pessimism:

> "To resist is not to be against, anymore, but to singularize," writes Suely Rolnik, reflecting on the changing meanings of artistic practice since the Great Refusal of the 1960s. "All and any acts of resistance are acts of creation and not acts of negation." [10] . . . The great theoretical swing of the past three decades, from critical negation to use value and subversive affirmation, has left "progressive" practices wide open to every form of cooptation and complicity. Despite the autopoetic processes that an installation like USE so brilliantly lets us see, the entire planet – Spaceship Earth – is prey to a resurgence of repressive authority, within the perfectly legible game of the capitalist world-economy. Berlusconi's Italy, where the project

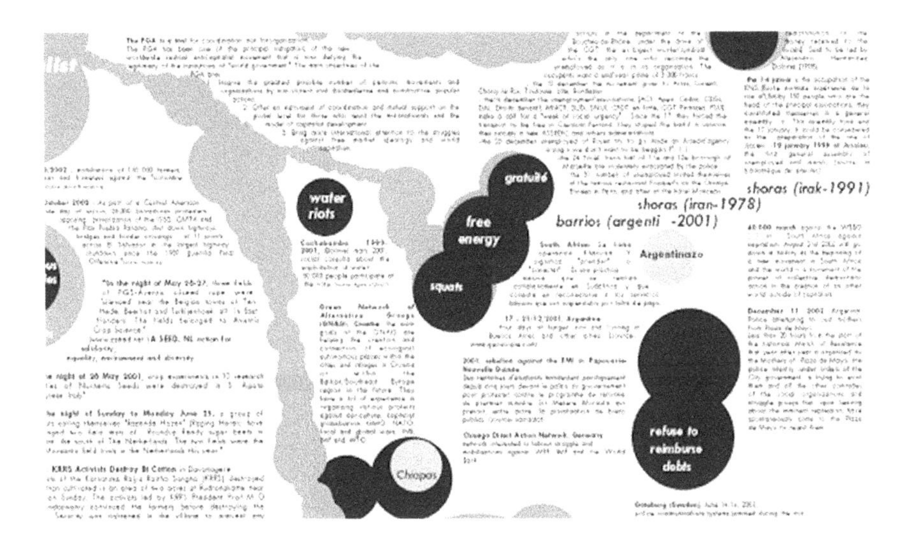

Figure 3.7 "Que se vayan todos". http://utangente.free.fr/2003/quesebayan.pdf

has been shown, is hardly an exception: and yet it is also one of the laboratories for new forms of political mobilization. Can we imagine artistic representations of self-organizing processes, in open confrontation with the economic game?

Deleuze himself not only coined the term "new cartography" to describe Foucault, but he similarly pushed Foucault's genealogy of power into a more contemporary frame in which he too described a kind of dystopic "society of control." This was a society in which:

> We are in a generalized crisis in relation to all the environments of enclosure – prison, hospital, factory, school, family. The family is an "interior," in crisis like all other interiors – scholarly, professional, etc. The administrations in charge never cease announcing supposedly necessary reforms: to reform schools, to reform industries, hospitals, the armed forces, prisons. But everyone knows that these institutions are finished, whatever the length of their expiration periods. It's only a matter of administering their last rites and of keeping people employed until the installation of the new forces knocking at the door. These are the societies of control, which are in the process of replacing disciplinary societies. "Control" is the name Burroughs proposes as a term for the new monster, one that Foucault recognizes as our immediate future.
>
> (Deleuze http://www.nadir.org/nadir/archiv/netzkritiksocietyofcontrol.html)

Deleuze was, however, very clear about the politics that flows from this understanding:

> There is no need to invoke the extraordinary pharmaceutical productions, the molecular engineering, the genetic manipulations, although these are slated to enter the new process. There is no need to ask which is the toughest regime, for it's within each of them that liberating and enslaving forces confront one another. For example, in the crisis of the hospital as environment of enclosure, neighborhood clinics, hospices, and day care could at first express new freedom, but they could participate as well in mechanisms of control that are equal to the harshest of confinements. There is no need to fear or hope, but only to look for new weapons.
>
> (http://www.nadir.org/nadir/ archiv/netzkritik/societyofcontrol.html)

The new cartography points precisely towards this search for new weapons within contemporary European social movements. Here a new generation of technically trained, computer-savvy activists is hooking up with what Graeber (2002b) has called the "new anarchists" to produce what we have called the new cartographers. To what extent this "cartographic turn" can be seen as a general trend or a minority stream within the movement of movements is not, at this point, the interesting question for us. What is fascinating in these cases is how a more rhizomatic politics is enacted, aimed at the analysis of the bio-political structures and networks emerging to shape, map, and bound European lives, and the possibilities they portend for other productions of subjectivity and modes of managing the care of the self.

In this alternative political logic, performance and chance have become important tactics and indeterminacy is emerging as a governing epistemology. The point is not only to understand effects and thereby to be able to monitor and adjust practices to create more effective and broadly based mobilizations. Equally important is the ability to create new imaginaries, tools and spaces here and now, that may be able to spread like memes through channels of formal and informal movement communication but whose longer term effects may be difficult to ascertain beyond the subjective effects they have had on their participants. It isn't that longer term effects are not important. They are. But these autonomous political practices and the social movements that deploy them do not want to have to depend only on logics of collective action that mobilizes force strategically and rationally in such a fashion.

Above all the new cartographers are experimentalists, willing to produce installations and events and allow them to be carried away, drift away, or be destroyed without obvious trace. Theirs is above all a politics of hope, not only that other worlds are possible but that they can be built in the process of living, rather than as a deferred moment of conjuncture when the forces align that may allow for broader mobilization. And it is a questioning ontology:

> We wonder about the several methodologies to develop those maps, as individuals or as group creations. As pieces of activist social communication, what

> are their repercussions? Are they creating new forms of mediatical interferences inside the mass mediascape? Are they able to help in the appropriation of information to be turned in collective action? Are maps/visuals a subject for specialists and experts? Even if it is obvious that there are real difficulties to develop, program, read and analyze maps, in which ways do people and groups involved in making maps do try to avoid those problematics? Those last questions raise also the cartographic methodologies to work with diverse variety of publics on social and political thematics. Developing maps and cartographies can also mean to develop collective methodologies of empowerment. Maps can mean insertion, immersion, exchange, conversations, several levels of sociabilities, the rise of geo-poetic tools, intimate maps, to be produced by kids, young ones, immigrants, women, communities of neighbours etc. How to develop in group collective tactical maps? where is the border between individual and activism work? where are the technical possibilities to develop decentralized maps from several groups perspectives etc?
>
> (http://www.euromovements.info/english/news3.htm)

Perhaps this is precisely the impulse behind the articulation of social movements, art, and new cartographies, in which representations are not to be defined "as the refusal of the commodified object and the specialized art system, but as an active signage pointing to the outside world, conceived as an expanded field for experimental practices of intimacy, expression and collaboration – indeed, for the transformation of social reality" (Holmes 2003).

Notes

1 This is evocative of what Lefebvre (1962) seems to admire of the Paris Commune and what he later called for as a focus on spatial conflict as a primary site for class conflict and revolutionary transformation. Precisely the creation of mass utopian spaces and fissures with dominated abstract spaces is what he saw as "correct" in the Commune and what he saw as completely absent from the mainstream communism of his time.

2 See Lefebvre's (1991) call to create "maximally" different spaces in order to replace capitalism's "abstract space" and its colonization of everyday life with his own utopia, "differential space," or spaces where substantive differences in ways of being could emerge and flourish.

4 A comprehensive list of these and like projects would be impossible, but to get sense of this diffusion and different sorts of activist cartography see Mogel and Bhagat (2007) and the upcoming traveling map archive "We Are Here" part of the exhibit "Experimental Geography" (http://danieltucker.wordpress.com/2008/01/16/cfp-for-map-archive/ or http://www.ici-exhibitions.org/exhibitions/experimental/experimental.htm.)

5 For example, see work on GeoSpatial Intelligence especially the recent talk by Todd Hughes on the "Mapping Revolution" (http://www.darpa.mil/DARPATech2007/proceedings/dt07-ixo-hughes-mapping.pdf).

6 These efforts have resulted in other collectives experimenting with map-making as a form of intervention and have even helped create a spin-off series of political seminars discussing new geo-economic and geopolitical changes through the practice and metaphor of mapping. These seminars include periodic engagement and dialogue with university scholars, such as the recent seminar '*Continental Drift*' in New York with David Harvey (see http://www.16beavergroup.org/drift/).

7 See the People's Global Action website (http://www.nadir.org/nadir/initiativ/agp/mayday1.htm) for more information.

8 Bureau d'Études Maps (selected), http://utangente.free.fr/anewpages/cartes.html
Capitalism Archive
"Governing by Networks" – http://utangente.free.fr/2003/governingbynetworks.pdf
"L"Industrie de Normalisation Europeene" http://utangente.free.fr/2003/europe A3.pdf
"World Government: Barclays plc" – http://utangente.free.fr/2003/barclays.pdf
"Infowar/Psychic war" – http://utangente.free.fr/2003/media2003.pdf
"Lagardere: Chronique de Guerre" – http://utangente.free.fr/2003/lagardere.pdf
"Europe" – http://utangente.free.fr/2003/europeA3.pdf
"Media Skills" – http://utangente.free.fr/2003/medias.pdf
"Rothschild" – http://utangente.free.fr/2003/rothschild.pdf
"Psywar" – http://utangente.free.fr/2003/PSYWAR.pdf
"Refuse the Bio-Police" – http://utangente.free.fr/biopolice/biopolice_intro.pdf
Autonomous Knowledges and Powers
"Que se Vayan Todos" – http://utangente.free.fr/2003/quesebayan.pdf

9 Bureau d'Études has additionally helped to establish the *Université Tangente* (UT), a free web resource and networking space for critical reflection and research regarding global transformations, radical activism, and political subjectivity. In turn, this has inspired other similar autonomous projects engaging with spatial practices and mapping (see http://www.edu-factory.org/mappa.html).

10 In this sense, the work of the Bureau d'Études can be seen as an important contribution to the often overused trope of "flow" as a way to think about globalization. While work such as Appadurai's (1996) writings on *scapes* has been helpful in early conceptualizations of the spatialities of globalization, the notion of "flow" has become an often too easy way to think about and naturalize the global (Thomas Friedmann's recent books *The Lexus and the Olive Tree* and *The World is Flat* come to mind here). Bureau d'Études helps by attempting to trace out the "flows" as multiple nexuses of complex institutional, legal, economic, or political relations.

11 At the encounter of Fadaiat 2005 (a gathering dedicated to rethinking forms of activism and grassroots communications in southern Spain and northern Morocco) this was one of the main calls launched from the workshop on "Tactical Cartographies", where Bureau d'Études among others participated. Since then different projects have been initiated, e.g. the Mapping Contemporary Capitalism project by *MUTE* magazine and others, a draft outline for a program by Bureau d'Études themselves, and at least two different attempts at designing open source mapping software between hackers and activists, Mapomaitx and Car Tac.

12 Hackitectura also has a complex relation to institutional academic and artistic spaces. While individual members may be working as professors or design consultants, Hackitectura's collective projects often occur tangentially to the university and the large studio.

13 See also: plug & play social event, propuesta para atenas 2003/agenciamiento de espacios públicos, redes tecnológicas y redes sociales/wewearbuildings.cc 2002, detalle/versión completa en: http://www.hackitectura.net/osfavelados/athens_web/plug_play_athens.html

14 Holmes uses in particular a map called "Centers and Peripheries in the World" that opens up the book *La Mondialisation du Capital* by Francois Chesnais (1997, Paris: Syros). Interestingly, Holmes notes that Castells draws on Chesnais several times for his analysis in *The Rise of the Network Society*. The map is a curious adaptation of the dymaxion image – known for breaking North–South dichotomies – superimposed with a stark center–periphery representation of power (Holmes 2004b: 2–3, see also https://pzwart.wdka.hro.nl/mdr/pubsfolder/bhimages/flowmaps/chesnaismap.jpg.

4 From surfaces to networks

Barney Warf

One of the most famous debates in the intellectual history of time and space took place in the seventeenth century between Isaac Newton and Gottfried Leibniz. For Newton, greatly influenced by the invention of the clock, space was like time: If the clock showed that time existed independently of events, then the same was true of space. Newton viewed time and space as abstract, absolute entities that existed independently of their measurement, ie., their existence was absolute, for their reality remained real regardless of whatever they contained or how they were measured. He argued in 1687, for example, that "Absolute, true, and mathematical time, of itself, and from its own nature, flows equally without relation to anything external" (quoted in Kern 1983: 11). Leibniz, in contrast, disliked the primacy of geometry in Cartesian thought, the implicit priority it assigned to space over time. Leibniz held that time and space were relational rather than absolute in nature, i.e., comprehensible only through frames of interpretation: distance, for example, could only be understood as the space between two or more objects situated in space. Space and time, therefore, had no independent existence in and of themselves, but were derivative of how we measured them. Eventually, for reasons having little to do with inherent intellectual merit and much to do with the emergence of early capitalist modernity, Newton's view triumphed decisively and Leibniz's relational space was "resoundingly defeated in the Enlightenment" (Smith 2003: 12).

This chapter is predicated on the argument that metaphors are a critical means by which we understand and appreciate the importance of space and spatiality in social life. Metaphors are a means of bridging the known and the unknown, and reveal much about a society's assumptions and cultural logics, what it holds to be important or normal, or not (Barnes 1991). The focus here is on two radically different metaphors of spatial relations, surfaces and networks, one historic and the other contemporary, which have underpinned popular and intellectual conceptions of geography in a multitude of ways. It begins by examining the rise of surfaces under early modern capitalism, the roles they played in fostering a distinct understanding of space conducive to the circuits of capital accumulation then stretching across the planet, the rise of the nation-state, and the certainty of visual knowledge or "scopic regime" (Gregory 1994) that they sustained. Second, it turns to the dynamics of postmodern, globalized capitalism, a world in which space plays a very different role, one in which nation-states have declined in significance, and argues that surfaces have been displaced by a new metaphor more suitable to the contemporary age: networks. Drawing on the works of several poststructuralists, it offers several examples by which relational space may be conceived and portrayed in terms more appropriate to the current historical moment. In both cases, the metaphors of space do much more than simply reflect social circumstances; rather, they actively contribute to their making.

Surfaces: making space visible to the ocularcentric ego

The global expansion of capitalism that began in the sixteenth century brought with it innumerable changes in economic, political, cultural, and ideological relations. The trading and colonial empires that flourished across the planet brought together vast realms of the planet under Western political, economic, and ideological domination. Global conquest also initiated a new understanding of the world among the elites of Europe, or what Pratt (1992) calls an incipient planetary consciousness, an understanding of the globe as a unified entity in which localities were deeply intertwined. It is Eurocentric to portray this process as simply one of reaching out across the surface of the earth; as Massey (2005: 4) notes, "Conceiving of space as in the voyages of discovery, as something to be crossed and maybe conquered . . . makes space seem like a surface: continuous and given" rather than as a mutable social production.

A central figure in the Renaissance reconceptualization of space was the mathematician and philosopher René Descartes. Among his other accomplishments, he may be regarded as the founder of modern ocularcentrism, the epistemological standpoint of early modernity that subscribed to the notion of a detached, objective observer capable of a "god's eye" view of the world. Descartes proposed an explanatory model centered on what has come to be known as the Cartesian cogito: a disembodied, rational mind without distinct social or spatial roots or location (but implicitly male and white). Cartesian rationalism was predicated on the sharp distinction between the inner reality of the mind and the outer reality of objects; the latter could only be brought into the former, rationally at least, through a neutral, disembodied gaze situated above space and time. Such a perspective presumes that each person is an undivided, autonomous, rational subject with clear boundaries between "inside" and "outside," i.e., between self and other, body and mind. In geographic terms, ocularcentrism equates perspective with the abstract subject's mapping of space. With Descartes's cogito, vision and thought became funneled into a spectator's view of the world, one that rendered the emerging surfaces of modernity visible and measurable and simultaneously rendered the viewer bodyless and placeless. Illumination was conceived to be a process of rationalization, of bringing the environment into consciousness through the modality of vision. Cosgrove (1999: 18) observes that "Modernity is distinguished by its concern with the human eye's physical capacity to register and to visualize materiality at every scale." The perspective had deep roots in Western history: Ó Tuathail (1996: 70) posits that "What was initiated in Greek philosophy was augmented by the innovations of perspectivalism and Cartesianism. Perspectivalist vision made a single sovereign eye the center of the visible world." Gregory (1994) likewise maintains that the early modern knowledge/power configuration – an historically specific scopic regime – reproduced reality in the form of the "world-as-exhibition." Cartesianism was thus simultaneously a model of knowledge and of the "individual," i.e., an observer devoid of social origins and consequences.

Co-catalytic with the Cartesian model of the human subject was the geometric view of space that it suggested; the ascendance of vision as a criterion for truth merged Euclidean geometry with the notion of a detached observer (Hillis 1999). This worldview had powerful social and material consequences. Cosgrove (1988: 256) notes, for example, that "in late Renaissance Italy not only was geometry fundamental to practical activities like cartography, land survey, civil engineering and architecture, but it lay at the heart of a widely-accepted neo-platonic cosmology." Ideologically, this process led to the mathematicization of the sciences, the search for a single set of universal laws, and an enormously powerful scientific worldview that greatly expedited Europe's technological prowess. As this perspective gained currency throughout Europe, the multiple vantage points in art or literature typical of the medieval world were steadily displaced by a single disembodied, omniscient, and panopticonic eye. Rather than the convoluted visual and aural worlds central to the medieval world, Renaissance thought emphasized homogeneous, ordered visual fields. "It was this uniform, infinite, isotropic space that differentiated the dominant modern world view from its various predecessors" (Jay 1993: 57). That this view arose precisely during the birth of modern science was hardly coincidental: as Jay (1993: 52) put it, like vision, space "functioned in a similar way for the new scientific order. In both cases space was robbed of substantive meaningfulness to become an ordered, uniform system of abstract linear coordinates." Thus, it was no accident that the Cartesian model arose in tandem with capitalism, colonialism, and modern science (Kirby 1996). The ascendancy of ocularcentrism also initiated the long-standing Western practice of emphasizing the temporal over the spatial. Ó Tuathail (1996: 24) argues that "the privileging of the sense of sight in systems of knowledge constructed around the idea of Cartesian perspectivalism promoted the simultaneous and synchronic over the historical and diachronic in the explanation and elaboration of knowledge."

The Renaissance rationalization of space was manifested in the explosion of cartography, which, as Harley (1989) stressed, replicated the assumption of an all-seeing, invisible creator cloaked in the mantle of objectivity. The rise of modern cartography represented a shift in vision from local topologies to a fine, spatially referenced, spherical earth, a homogeneous graticule of latitude and longitude. The epistemology of cartography – its ocularcentrism, its purportedly scientific rendition of space – served a vital social function. Yet far from constituting a detached, objective viewpoint from nowhere, a view that reduces map-making to a *technical* process, cartography was a *social* process deeply wrapped up in the complex political dynamics of colonialism. Under the expansion of colonial empires, the need to represent distant places – to make them present for those who were not there – rose exponentially. Renaissance cartography thus effectively consisted of the "geographing" of remote regions to facilitate their control. After all, in order to get to, conquer, govern, and administer their colonies, the Europeans first had to know them spatially. The grid formed by latitude and longitude was one of several such systems deployed worldwide to facilitate the exchange networks of early modern capitalism, making space smooth, fungible, and comprehensible by imposing order on an otherwise chaotic environment. For European navigators, this move entailed smoothing space by reducing it to distance, rendering the oceans navigable, and ordering the multitude of world's places within a comprehensible schema. The projection of Western power across the globe necessitated a Cartesian conceptualization of space as one that could be easily crossed, a function well performed by the cartographic graticule. Inserting various places in all their unique, messy complexity into a global skein of meridians and longitudes positioned innumerable locales into a single, unified, coherent, and panopticonic understanding of the world designed by Europeans, for Europeans, allowing different locations to be compared and normalized within an affirmation of a god-like view over Cartesian space at the global level. Colonial mapping was thus not simply a tool for administration, but equally importantly, a validation of Enlightenment science and central part of the colonial spatial order: mapping offered both symbolic and practical mastery over space.

Thus, the discourses of space did far more than passively represent entities that existed before them, but were actively a part of *producing* that very geography; they were, in short, simultaneously reflective and constitutive of the reality they represented. By subjecting the planet's diverse people to the conceptual lens of Western modernist rationality, cartography enfolded the world within a particular Western way of understanding, one that erected reality as a picture to be gazed upon from a distance, a totalized actuality that was ordered and structured, i.e., Gregory's (1994) "world-as-exhibition." O'Tuathail (1996: 53) notes that this process mirrored the ascending ocularcentrism of the age:

> By gathering, codifying, and disciplining the heterogeneity of the world's geography into the categories of Western thought, a decidable, measured, and homogeneous world of geographical objects, attributes, and patterns is made visible, produced. The geopolitical gaze triangulates the world political map from a Western imperial vantage point, measures it using Western conceptual systems of identity/difference, and records it in order to bring it within the scope of Western imaginings.

A parallel transformation occurred within the visual arts, in which bourgeois values became increasingly hegemonic (Cosgrove 1984). The key discovery in this regard was the invention of linear perspective, first demonstrated by Filippo Brunelleschi in 1425, which involved the ability to represent three dimensions on a two-dimensional canvas. Thus, "Linear perspective vision was a fifthteenth-century artistic invention for representing three-dimensional depth on the two-dimensional canvas. It was a geometrization of vision which began as an invention and became a convention, a cultural habit of mind" (Romanyshyn 1993: 349). Mumford (1934: 20) noted that "Perspective turned the symbolic relation of objects into a visual relation: the visual in turn became a quantitative relation. In the new picture of the world, size meant not human or divine importance, but distance." As Johnson (2002: 118) argues, the "replacement of aspective art by perspective art was one of the greatest steps forward in human civilization." Like cartography, perspective painting served a social purpose and a specific, hegemonic power interest, and came into being just as Florence came under the panopticonic gaze of the Medici aristocracy (Edgerton 1975).

Simultaneously, the process of printing deepened and reinforced the emerging European ocularcentrism. Printing was the first major step in the mechanization of communication, and accelerated the diffusion of information by packaging it conveniently, in the process democratizing books. This process undermined the centrality of the clergy in the production of knowledge: books, unlike hand-written monastic manuscripts, gave their audiences identical copies to read, experience, and discuss, and made censorship more difficult as well. Printing thus broke the monopoly on learning held by monks and fostered the growth of a lay intelligentsia. Eisenstein (1979) demonstrated the enormous power of printing in diffusing knowledge and mass literacy, facilitating the Renaissance, the Protestant Reformation, European expansionism, and the rise of modern capitalism and science. Printing in vernacular languages began to undermine the hegemony of Latin, establishing local tongues as the basis of emergent national identities and imagined communities. Anderson (1983) famously argued that nationalism co-evolved with the growth of print-based culture once vernacular languages became the norm of printed communications, as the printing press connected disparate populations spread over wide geographical areas. This process had enormous effects on the social and spatial structure of language. As the market for books in Latin became gradually saturated, vernacular languages became increasingly common and popular. Moreover, this process led to the very idea of a fixed point of view, a foundational part of the Cartesian metaphysic that underpinned both modern science and modern perspectives on time and space.

Printing did more than simply accelerate the dissemination of knowledge, ideas, and information, it also reinforced the emerging ocularcentrism of early modernity. As Jay (1993) and Jenks (1995) noted, the rise of printing, the reliance on the written word for communication, and the use of the telescope and microscope to bring the distant and the invisibly small into view all contributed to the tendency to equate seeing with knowing. Epistemologically, printing suggested that words were things, situating words in space far more than did writing and embedding language in the process of manufacturing, which in turn accelerated its commodification. The printing of maps began to accustom Europeans to visual, grid-based representations of territorial order, helping to establish abstract space as the dominant model of the early modern period. Thus, printing utilized the spatial organization of knowledge widely and effectively, generating visual surfaces with abundant and intense meanings, with enormous consequences for human perceptions of space and time.

Finally, the construction of modernist geographies in the form of surfaces rested heavily on the rise of the nation-state: Smith (2003: 142) argues "The genesis of national states as a system for organizing the world's political economy provided an eighteenth-century 'spatial fix' for specific economic dilemmas of emergent capitalism." The international system legitimated by the Treaty of Westphalia in 1648 underscored the centrality of the nation-state to the early modern world system, a world of absolute spaces and explicit, non-overlapping boundaries. Such a geopolitical structure was unprecedented: "The modern state system of territorially fixed and mutually exclusive sovereignties is an historically unique form of spatial organization" (Anderson 1996: 140).

The shift in scale from the city-state to the "power-container" of the nation-state (Giddens 1984) reflected a large variety of factors, including: the intensified commodification of land and labor; the centralization of law enforcement, particularly regarding property rights; printing and explicit codes of law; the diffusion of paper money as a medium of exchange; apparatuses of taxation, surveillance, and documentation such as the census; the judiciary and penal systems; and gradual improvements in transportation such as turnpikes and canals. Within nation-states, banking systems created homogeneous financial spaces in which the cost of capital was almost totally invariant. Communications systems such as the postal service allowed more effective governance (not necessarily democratic), and the ability to mobilize the masses during emergencies. Military drafts socialized young men from disparate backgrounds into a national identity. Mass literacy, newspapers, and the ideology of nationalism also contributed to the homogenization of culture that turned feudal societies into nation-states. Foucault (1972) stressed that under the disciplinary logic of modernity, vision became supervision: it lost the benign status of the detached observer and became a means of enforcement and surveillance. He argued that subtle social mechanisms of control – the police, the medical system, education, etc. – extended the power of the omnipotent sovereign to produce subjects who self-monitored their behavior, conforming to the taken-for-granted notions instilled in them from birth. Nationalism also transformed abstract space into a territory imbued with selective interpretations of local history, a homeplace that fused the immemorial past with the future destiny of its people. By the early nineteenth century, increasingly standardized public education systems played a central role in linking individual identities to the state, i.e., raising the scalar level at which people defined themselves and one another. In producing the citizen, the nation-state also constructed moral geographies of similarity and difference, inclusion and exclusion, which sharply distinguished "us" from "them," amplifying the differences between the community of insiders and foreign outsiders.

This homogenization was perfectly in keeping with the Cartesian view of space: as with linear perspective in painting, the nation-state in geopolitics came to be defined from a single, fixed viewpoint (Ruggie 1993). "As containers of a fledgling modernity, the expansionist new monarchies of the sixteenth century were slowly, unevenly, and erratically (depending on the state in question) imposing a general perspectivalist vision of space and a neutral conception of time upon the territories they incorporated and annexed" (Ó Tuathail 1996: 12). In contrast to feudal empires, which often had diffuse boundaries, the nation-state was predicated upon a view of geography as Euclidean, a "horizontal order of coexistent places that could be sharply delimited and compartimentalized from each other" (Ó Tuathail 1996: 4).

Surfaces as the predominant form through which modern space was constituted ontologically and understood epistemologically continued to have a long and enduring emphasis in the twentieth century. The rise of logical positivism in the late nineteenth century added a aura of scientism to this view, mathematicizing it with the disciplines concerned with space such as geography and urban planning in the forms of isotropic planes, surfaces in which the distribution of social features is evenly distributed, e.g., the von Thunen model (see Isard 1972). Thus, approaches such as Central Place Theory continued a long tradition of ocularcentrism (Gregory 1994). The emphasis on transport costs, as in Weberian location theory, reflected their role in the location decisions of firms. Similarly, urban space was largely depicted in terms of surfaces of differential rent and commuting fields. During the post-World War II economic boom, in which markets were largely national in scope, transport costs were relatively high, knowledge and information were comparatively small proportion of costs of production, and the Fordist social contract held sway, the view that geography consisted of neatly nested surfaces of minimal heterogeneity was both utilitarian and expedient. Places in this conception tended to be viewed as isolated and static, and space in general was relegated to a passive role, a mere reflection of economic logic, not an actively constituent part.

The collapse of the surfaces metaphor occurred not because of some abstract shift in ideas, but because the world that sustained it drew to a close in the late twentieth century. The growth of producer services and "post-industrial society" witnessed the steady decline in transportation costs and the rise in importance of information as the primary input and output of the division of labor. The steady shift from national to global markets likewise undermined the relevance of models predicated on isolated localities. Such changes were reflected in different meanings assigned to distance, in which topological relations became increasingly important. As the social and material conditions of modernism gave way to the convoluted geographies of globalized post-Fordism, the view of geography as simple surfaces, correspondingly, began to yield to new understandings that took a very different spatial metaphor as their point of departure.

Networks: retheorizing space in the age of flexible globalization

The profound global crisis in Fordism that ended the post-war boom in the 1970s was accompanied by numerous birth pangs of a new global economic and social order, including: the petroshocks; prolonged deindustrialization in the West; the rise of the Newly Industrializing Countries, particularly in East Asia; the end of the Bretton-Woods era in 1973 and subsequent shift to floating exchange rates;

the growth of third world debt, largely driven by recycled petrodollars; the emergence of "flexible" specialization, the microelectronics revolution, and computerized production technologies; and the rapid, sustained growth of financial and producer services (Dicken 2007). This collection of events may be viewed as moments in the transition into postmodern "flexible" and "disorganized" capitalism (Lash and Urry 1987). Similarly, Bauman (2000) differentiates "heavy" modernity, a period lasting from the Renaissance to the late twentieth century, from the "light modernity" of contemporary capitalism, one centered on mobile rather than fixed capital, speed rather than power, instantaneity rather than duration, software rather than hardware, individual rather than collective struggle, consumption rather production.

Globalization has arguably become the defining process of the contemporary world. This process extends deep into the historical geography of capitalism; Smith (2003), for example, argues that the close of the colonial era at the end of the nineteenth century was marked by the need to view space in relative rather than absolute terms. While there have been many episodes of globalization in the past, the current round, which began in the 1970s, has been particularly rapid and penetrating. So extensive has the acceleration in the scope, magnitude, and velocity of international transactions become that few analyses of local events and processes can afford to ignore it.

Central to the rise of globalized, postmodern capitalism was the construction of a vast web of telecommunications networks, which were crucial to the hegemony of increasingly information-intensive capitalism (Warf 1995; Schiller 1999). The core of the global telecommunications infrastructure is a seamless network of fiber optics lines (Graham 1999). Because the implementation of fiber lines reflects the powerful vested interests of international capital, these systems may be seen quite literally as "power-geometries" (Massey 1993) that ground the space of flows within concrete material and spatial contexts. Financial services firms were at the forefront of the construction of high-capacity fiber networks in large part because they facilitated the use of electronic funds transfer systems, which comprise the nervous system of the international financial economy, allowing banks to move capital around a moment's notice, arbitrage interest rate differentials, take advantage of favorable exchange rates, and avoid political unrest. Conventional geographic preoccupations with proximity mean little in this case, for the topologies of global finance are formed and deformed at the speed of light in ways that the language of location theory cannot capture.

The largest and most famous of the world's telecommunications networks is undoubtedly the internet, an unregulated electronic network connecting an estimated one billion people in more than 180 countries in 2007. It is arguably the quintessential symbol of postmodern capitalism: electronic, globalized, and rapidly evolving. From its military origins in the USA in the 1960s, the internet emerged upon a global scale through the integration of existing telephone, fiber optic, and satellite systems (Rosenzweig 1998). Spurred by declining prices of services and equipment, the internet grew worldwide at stupendous rates, making it the most rapidly diffusing technology in history. The internet also annihilates distance to an unprecedented degree: distant locales are now just a mouse-click away, allowing users to transcend distance virtually instantaneously, generating total time-space compression.

Although their potential impacts are often exaggerated, digital networks clearly have substantial, if largely unanticipated, effects upon the social and economic fabric over time (Kitchin 1998). Gregory (1994: 98) maintains that "ever-extending areas of social life are being wired into a vast postmodern hyperspace, an electronic inscription of the cultural logic of late capitalism." E-commerce transformed the internet from a communications to a commercial system, widening commodity chains and accelerating the pace of customer orders, procurement, production, and product delivery. As virtual reality and "real" reality have become ever more tightly interwoven, the digital world has exerted a rapidly increasing influence over the social fabric, to the point where distinguishing between these two domains no longer seems helpful. Software, for example, through the mutually constitutive relationship it enjoys with territory, enables space to unfold in multiple ways, such as when it is used to monitor and control automobile and airline traffic, retail trade, and monitor spaces through closed-circuit television (Dodge and Kitchin 2004, 2005a, 2005b, 2007). So deeply have email, the World Wide Web, and e-commerce penetrated various domains of social life that simple dichotomies such as offline/online are insufficient to account for how the virtual and real worlds are interpenetrated. Cyberspace may even rework the nature of urban space itself: whereas the modern city was characterized by stationary architecture and moving masses, the virtual city is one in which the masses sit still while the cinematic space of flows swirls around them (Crang 2000). Indeed, in a socio-psychological sense, cyberspace may allow for the reconstruction of "communities without propinquity," groups of users who share common interests but not physical proximity, although the ability of virtual communities to substitute for face-to-face ones is debatable. The implications of this process are sobering. As Graham and Aurigi (1997: 26) note, "Large cities, based, in the past, largely on face-to-face exchange in public spaces, are dissolving and fragmenting into webs of indirect, specialized relationships." More generally, cheap, instantaneous, and ubiquitous communications have made the notion of place as a discreet, bounded entity increasingly problematic by allowing people to be in several places simultaneously. In this context, absolute distance is much less relevant than relative distance, i.e., space measured in terms of time, cost, and social access.

The ability to transmit vast quantities of information in real time over the planet is crucial to what Schiller (1999) calls digital capitalism. No large corporation could operate today in multiple national markets simultaneously, coordinating the activities of thousands of employees within highly specialized corporate divisions of labor, without access to sophisticated channels of communications. Telecommunication networks also gave many firms markedly greater freedom over their locational choices. Transnational corporations – the networks' primary users and beneficiaries – rely heavily upon such networks to coordinate and monitor international transactions, which can be monitored across the globe as easily as if they were in the same building. In dramatically reducing the circulation time of capital, telecommunications linked far-flung places together, creating a geography without transport costs. The effectiveness of national controls was markedly reduced by the ease with which hypermobile financial capital moves through global markets. Thus, as large sums of funds flowed with mounting rapidity across national borders, Keynesian monetary monetary policies became increasingly ineffective. In a Fordist world system, national monetary controls over exchange, interest, and inflation rates are essential; in the globalized post-Fordist system, however, those same national regulations become a drag on competitiveness.

The spatial scale at which globalization operates suggests that national institutions and processes, which dominated throughout the nineteenth and early twentieth centuries, may have reached the limits of their effectiveness. The steady reduction in the significance of borders leads inexorably, if unevenly, to a seamlessly integrated world, a process that is as much political in its construction (e.g. via trade agreements) as it is economic. Although the argument that the nation-state is disappearing has been advanced in occasionally exaggerated and simplistic terms (e.g. Ohmae 1990; Gergen 1991), the enormous tsunami of postmodern globalization has fundamentally undermined the territorial order of distinct, mutually exclusive, sovereign states that has underpinned the international order since the Westphalian Peace of 1648. The decline of the nation-state has been fueled, among other things, by the rising significance of global problems (e.g., global warming), the threat posed by subnational tribal conflicts and transnational ideologies (e.g., Muslim fundamentalism), and mounting international trade and integrated financial markets. In this light, the nation-state is being eroded simultaneously "from above," i.e., by the growing power of transnational public and private institutions (e.g., the IMF, World Bank, and World Trade Organization), as well as those working "from below," that is, regional authorities and local movements seeking to attract foreign capital or participate in global processes by bypassing their respective national states as well as NGOs (Anderson 1996). This transformation makes the simple difference between "inside" and "outside" increasingly problematic, rendering national borders ever more porous. Numerous observers have therefore suggested that the postmodern world order is becoming "unbundled" or "plurilateral" (e.g., Ruggie 1993; Elkins 1995; Gilpin 1987), that is to say, it is being displaced by a new territorial configuration not predicated on the nation-state. In some respects, the unbundling of territorial sovereignty represents a return to medieval structures of territoriality, in which the boundaries between the internal and external were messy and ill-defined; Lipschutz (1992), for example, describes the postmodern world order as something akin to the medieval geography of Europe, in which corporations replace the Christian church as the primary source of aerial integration.

However, it is simplistic to assume that globalization leads inevitably to the end of nation-states as they are currently constituted, replacing them with some mythical, seamless integrated market that embraces the entire planet. Globalization is always refracted through national policies (e.g., concerning labor, the environment), which is one reason it has spatially uneven impacts across the world. Capitalism is a system predicated on both markets and states, and the political geography of globalization is the interstate system (but not the nation-state), the existence of which is necessary in order for capital to play states and localities off against one another. While globalization clearly undermines the ahistorical reification of nation-states as fixed, eternal entities, it does not automatically spell the death of this institution. Because the global and the local are shot through with one another, or in Swyngedouw's (1997) term "glocalized," globalization is manifested differently in different places (Cox 1997), a view that dispels simplistic popular assertions that globalization is synonymous with homogenization devoid of geographic specificity.

Resulting directly from the rise of networked post-Fordist production systems, a fundamental reworking of the relations between capital and space took place, a change with massive political ramifications. In climbing out of the crisis of Fordism, capital replaced the Keynesian national "spatial fix" (Harvey 1989) with a highly fluid, globalized set of localities, or what Swyngedow (1989) theorizes as regions embedded in a volatile, postmodern "hyperspace." Neoliberalism insured that innumerable places throughout the world have been opened up to global capital to form an increasingly seamless, if as yet incompletely integrated, series of networks. Unimpeded by national restrictions undone by deregulation, capital roams the globe effortlessly, testimony to the worldwide decline in both technological and regulatory barriers. Globalization does not lead automatically to a homogenization of local landscapes, however, as the global and local become entwined in complex, contingent regional formations of capital and labor.

Concomitant with these changes, social scientists have searched for new frames of reference to make sense of the emerging geographies of centrality and peripherality unleashed by flexible globalization. While place remains an enduring feature of contemporary social life, to an ever-increasing extent the geographies of postmodernity are defined by mobilities, flows, and networks rather than isolated, discreet locales (Crang 2002; Larsen *et al.* 2006). Under the impetus of distance-crushing time-space compression, in Kirby's (1996: 71) words, "The distinctions between widely separated geographical entities lose their meaning as disparate sites themselves come together in a plastic network making proximity and separation relative and mutable." Location, in short, has become increasingly a matter of production and negotiation rather than being given *a priori*. If the surfaces that characterized the geographies of modernity have declined in utility, the spatial metaphor that has taken their place centers on networks, which Ruggie (1993: 141) likens to the "economic equivalent of relativity theory." Unlike the surfaces of modernity (e.g., the nation-state, urban commuting fields), postmodern society exhibits a "fibrous, thread-like, wiry, stringy, ropy, capillary character that is never captured by the notions of levels, layers, territories, spheres, categories, structures, or systems" (Paasi 2004: 541). Not coincidentally, this transformation occurred precisely when geography moved from being a stagnant conceptual backwater to one taken increasingly seriously by other disciplines. Contemporary globalization has undermined commonly held notions of Euclidean space by forming linkages among disparate producers and consumers intimately connected over vast distances through flows of capital and goods. Arjun Appadurai (1996) has argued that the postmodern world is essentially fractal, that the Euclidean boundaries and structures characteristic of modernity have melted away in the face of the white-hot fires of time-space compression. In Sheppard's (2002) view, global power-geometries cut across spatial scales to create "wormholes" that defy the traditional logic of economic geography (also see Brenner 1998).

Temporally, the rise of networked capitalism has been accompanied by accelerated rhythms of production and interaction, with important consequences for the meaning of spatiality, or as Luke and Ó Tuathail (1998: 72) put it, "the power of pace is outstripping the power of place." Nonetheless, simply noting the acceleration of acceleration, so to speak, ignores deeper, more meaningful issues as to who benefits (and why) and who loses out (and why). "The really serious question which is raised by speed-up, by 'the communications revolution' and by cyberspace, is not whether space will be annihilated but what kinds of multiplicities (patternings of uniqueness) and relations will be co-constructed with these new kinds of spatial configurations" (Massey 2005: 91). Even more radically, Deleuze and Guattari (1987) offered a "schizophrenic" reconceptualization of postmodern space, one centered on rapidity, movement, and constant flux without the usual co-ordinates of distance and direction. The metaphor of the rhizome, so different from the transparent spaces of Euclideanism, offers a view of space as inherently and interconnected sets of networks, non-hierarchical in nature, in which connections among locales rather than their absolute positionality is the dominant characteristic. For this reason, rhizomes have become a popular means to envision networks such as the internet.

Envisioning space as networks rather than surfaces has taken several forms in poststructuralist social science. While there are wide variations among them, all poststructuralist perspectives take seriously the role of power, language, and historical context, the contingent nature of social life, and the heterogeneity among and within social formations. Three perspectives are examined here in short vignettes designed to reveal geography as interconnected networks rather than a series of discreet locales: the space of flows, commodity chains, and actor-networks. All of them emphasize space as produced, not given; all focus on the relational characteristics of place rather than absolute location; in all of them, distance is defined functionally rather than in terms of physical length; and all transcend the traditional notion of spatial scale as fixed.

The space of flows

Castells (1996, 1997) distinguished earlier *information* societies, in which productivity was derived from access to energy and the manipulation of materials, from later *informational* societies that emerged in the late twentieth century, in which productivity is derived primarily from knowledge and information. In his reading, the time-space compression of postmodernism was manifested in the global "space of flows," including the three "layers" of transportation and communication infrastructure, the cities or nodes that occupy strategic locations within these, and the social spaces occupied by the global managerial class:

> Our societies are constructed around flows: flows of capital, flows of information, flows of technology, flows of organizational interactions, flows of images, sounds and symbols. Flows are not just one element of social organization: they are the expression of the processes dominating our economic, political, and symbolic life. . . . Thus, I propose the idea that there is a new spatial form characteristic of social practices that dominate and shape the network society: the space of flows. The space of flows is the material organization of time-sharing social practices that work through flows. By flows I understand purposeful, repetitive, programmable sequences of exchange and interaction between physically disjointed positions held by social actors.
> (1996: 412)

For Castells (1996: 470), this transformation is best understood through the metaphor of rhizomatic networks:

> Networks are appropriate instruments for a capitalist economy based on innovation, globalisation, and decentralised concentration; for work, workers and firms based on flexibility; for a culture of endless deconstruction and reconstruction; for a polity geared towards the instant processing of new values and public moods; for a social organisation aiming at the supersession of space and the annihilation of time.

He notes, for example, that while people live in places, postmodern power is manifested in the linkages among places, their interconnectedness, as personified by business executives shuttling among global cities and using the internet to weave complex geographies of knowledge invisible to almost all ordinary citizens. This process was largely driven by the needs of the transnational class of the powerful employed in information-intensive occupations; hence, he writes (1996: 415) that "Articulation of the elites, segmentation and disorganization of the masses seem to be the twin mechanisms of social domination in our societies." Flows thus consist of corporate and political elites crossing international space on transoceanic flights; the movements of capital through telecommunications networks; the diffusion of ideas through organizations stretched across ever-longer distances; the shipments of goods and energy via tankers, container ships, trucks, and railroads; and the growing mobility of workers themselves. In this light, the space of flows is a metaphor for the intense time-space compression of post-Fordist capitalism. Through the space of flows the global economy is coordinated in real time across vast distances, i.e., horizontally integrated chains rather than vertically integrated corporate hierarchies. In the process, it has given rise to a variety of new political formations, forms of identity, and spatial associations.

Flows by definition involve more than one place; hence, in a networked world, places have little meaning as isolated entities. Places are not locales as much as they are processes in which different types of activities are embedded and different forms of interconnection are established. As they become increasingly connected, the repercussions of actions in one area inevitably spiral out to shape other places, so that discreet boundaries have less and less significance as they are permeated with mounting ease. Decisions made by hedge fund managers in New York, for example, reach out to affect the lives of millions of people in locations as distant as East Asia. The space of flows wraps places into highly unevenly connected chains that center on connections among powerful elites, thus typically benefiting the wealthy at the expense of marginalized social groups. However, the global space of flows is hardly randomly distributed over the earth's surface: rather, it reflects and reinforces existing geographies of power concentrated within specific nodes and places, such as global cities, trade centers, financial hubs, and headquarter complexes. Thus, far from being abstract networks divorced from history, such systems are only made comprehensible by embedding them in history (Cresswell 2006).

Commodity chains

Another step in the theorization of space as networks is the notion of commodity chains (Gereffi and Korzeniewicz 1994; Dicken *et al.* 2001). Drawing from the earlier French tradition of *filières* or value chain analysis, commodity chains may be defined as networks of labor and production processes that give rise to a particular commodity from raw material to processing, delivery, and consumption. Commodity chains include flows of goods and information among different points or nodes, varying labor relations across the length of the chain, and different constellations of production and governance at each segment. Such tools are useful in dissecting, for example, the spatiality of transnational corporations. Under globalization, commodity chains became increasingly larger in scope and length, linking widespread places through an increasingly complex division of labor in which distant strangers became ever more reliant upon one another.

Gereffi (1996) distinguished between producer-driven and buyer-driven chains, in which the power to control transactions lay at different ends of the concatenated linkages. In overcoming the artificial separation between production and consumption, commodity chains help to expose the widespread commodity fetishism prevalent in advanced economies, exposing commodities, *à la* Marx, as more than just things but as embodiments of social processes and thus helping to expose the unequal power relations among places that lie in the creation of goods. Chains are hence simultaneously geographical, economic, political, and cultural. They lie at the boundaries of the tangible and the intangible, incorporating sets of meanings, or as Hartwick (1998: 424) elegantly put it, they "conjoin the representational with the geomaterial." Moreover, they are temporally situated, constantly fluctuating in location, composition, and length over time.

Actor-networks

A third moment in this line of reasoning is actor-network theory (Murdoch 1997, 1998; Law and Hassard 1999). Inspired by the work of Bruno Latour (1993) (and

Serres 1997; Serres and Latour 1995), actor-network theory incorporates sociological understandings of structuration theory (Giddens 1984) with a poststructuralist, French, social-constructivist philosophy of science (Bingham and Thrift 2000). Actor-network theory departs from the Enlightenment focus on dualities such as individual and society, people and nature, human and nonhuman, Western and non-Western, urban and rural, micro and macro, local and global. Rather, it takes as its point of departure the linkages among these categories rather than their essences as actors draw upon and combine them in various forms of hybridity.

Networks involve the mobilization of rules, resources, and power, including information, in order to accomplish tasks, creating a net of intended and unintended consequences that stretch across the spatio-temporal boundaries of the network. To maintain network functionality, actors must perform by being engaged with one another recursively, interpreting and translating one another's behavior. Actors and networks are thus twin, mutually presupposing aspects of one phenomenon, simultaneously enabling and constraining actions in time and space. Because actor-network theory strives to overcome the artificial boundaries between culture and nature (Latour 1993), actors in this socio-technical seamless "nature-culture" nexus need not be human, but may include inanimate objects such as books, papers, or computer systems (Bingham 1996; Murdoch 1997), which are necessary to the maintenance and operation of networks.

Thus, it is not simply actors in everyday life who constitute the primary focus here, but their relative positionality and powers within integrated systems of power and information that matters most. For example, Thrift and Leyshon (1994) employed actor-network theory to examine the dynamics of global capital markets as they are structured by firms, nation-states, the media, and telecommunications, all of which are deployed simultaneously to produce, transmit, and consume knowledge about markets and other actors. Such a perspective has helped to humanize even the most abstract of economic processes by revealing them to be the products of agents enmeshed in webs of power and meaning, not disembodied processes that operate independently of the people who create them (Law 1994). The strategy of embodiment goes a long way toward demythologizing teleological interpretations of globalization, which present it as natural and inevitable, and reveal global processes to be the contingent outcomes of decisions made by human actors tied up in networks that cross multiple spatial scales.

Such a view elides the conventional focus on spatial scale, for networks operate across many scales simultaneously, creating as Latour (1993: 121) put it, "an Ariadne's thread that allows us to pass with continuity from the local to the global, from the human to the nonhuman. It is the thread of networks of practices and instruments, of documents and translations." Likewise, Massey's (1999) well-received notion of power-geometries has called attention to the intertwined scales of the global, national, and local, refusing to see these as a simple hierarchy in which the global determines the local; the distinctions among these scales are as misleading as they are enlightening. Smith (1993) argues that scale is produced through and constitutive of social relationships, and Thrift (1995: 33) goes so far to claim "There is no such thing as scale." By forcing us to rethink how time and space are produced – that is, topologically rather than in terms of conventional

Cartesian and Kantian views of space that have dominated geography – actor-network theory becomes "a machine for waging war on Euclideanism" (Law, quoted in Murdoch 1998: 357).

Concluding thoughts

The rise of post-Fordist globalized networks of people, capital, goods, and ideas changed many theorists' view of space from the notion of absolute, Cartesian notion – static, fixed, and lying outside of society, or space as a container – to relative and relational space, space as socially constructed by people, and thus fluid, folded, twisted by chains, pleated, and unstable (Murdoch 2006). Thus, the rise of social constructivism and the view of space as networks are deeply intertwined: to see geography as produced, not given, is to view it as consisting of and constructed by chains of causality in which distance is relational, not absolute. Unlike surfaces, which have traditionally been portrayed as containers "outside" of society and thus "holding" it, networks explicitly admit to their human construction. As Mann (1986) insists, societies are never unitary, bounded states, but multiple, overlapping, intersecting, and contingent networks of power stretched unevenly across time and space. No social formation is thus a totally unified entity, but rather forms an open-ended lattice of relations; societies do not just occupy space, they manufacture networks. Such a notion is disconcerting to those accustomed to thinking of geography in Cartesian terms, i.e., as the smooth surface of a globe. As Latour (1993: 77) maintains, "How are we to gain access to networks, those beings whose topology is so odd and whose ontology is even more unusual, beings that possess both the capacity to produce both time and space?"

In this light, geography is not simply territorial, but something altogether different, more complex, and more interesting. Postmodern, poststructural theories such as commodity chains and actor-networks greatly accelerated the rise of relational views of space. In this context, Harvey (2006: 121–3) offers a useful definition of absolute, relative, and relational space:

> Absolute space is fixed and we record or plan events within its frame. This is the space of Newton and Descartes and it is usually represented as a pre-existing and immoveable grid amenable to standardized measurement and open to calculation. Geometrically it is the space of Euclid and therefore the space of all manner of cadastral mapping and engineering practices. . . . The relative notion of space is mainly associated with the name of Einstein and the non-Euclidean geometries that began to be constructed most systematically in the 19th century. Space is relative in the double sense: that there are multiple geometries from which to choose and that the spatial frame depends crucially upon what it is that is being relativized and by whom. . . . The relational concept of space is most often associated with the name of Leibniz who . . . objected vociferously to the absolute view of space and time so central to Newton's theories. His primary objection was

> theological. Newton made it seem as if even God was inside of absolute space and time rather than in command of spatio-temporality. By extension, the relational view of space holds there is no such thing as space or time outside of the processes that define them. . . . Processes do not occur *in* space but define their own spatial frame. (Italics in original.)

Thus, whereas Newton's emphasis on absolute space prevailed in the seventeenth century and dominated for the next three centuries, the last laugh belongs to Leibniz.

Poststructural geography, in emphasizing the embodied nature of social life as situated practices, drew attention away from space as an inert container and toward interconnected sets of places as manifolds that are continuously folded and pleated, stretched, distorted, and shredded. This perspective has served to underscore how, unlike traditional chorology, with its emphasis on static places, relational geographies are always dynamic, incomplete, forever coming into being, and perpetually in flux, giving rise to ever-changing patterns of centrality and peripherality. Geography consists of the contingent networks or power-geometries generated by social interaction rather than a homogeneous plane that pre-exists coherent, well ordered societies. In short, space is emergent rather than existing *a priori*, it is composed of relations rather than structures. Relational space "is seen as an undulating landscape in which the linkages established in networks draw some locations together while at the same time pushing others further apart" (Murdoch 2006: 86). Taken to the extreme, Doel (1999) follows Deleuze and Guattari in emphasizing the origami-like nature of space as it is repeatedly folded and refolded, fissured, cracked, and fractalized through a series of difference-producing repetitions.

Massey (1993) criticizes notions that maintain place is an island of stability in the constantly shifting oceans of capitalist change, arguing that such a characterization is reactionary. Rather, she promotes a progressive sense of place that links places to other places, a view in which places constantly change, producing and receiving changes through their interactions with one another. A relational politics of place calls into question easy distinctions like inside/outside, near/far, space/place, and global/local, artificial differentiations that are always embedded in each other and mutually constituted. She argues passionately (2005) that Cartesian conceptions of space as a passive surface inevitably de-emphasize the temporal flux that is always an inherent part of geographies, and simultaneously create a false dichotomy between the local and the global. As an alternative, she suggests three maxims: 1) that space be seen as the product of interrelations, i.e., of embedded social practices in which identities and human ties are co-constituted; 2) that space be understood as the sphere of multiple possibilities, i.e., as a contingent simultaneity of heterogeneous historical trajectories; and 3) that space must be conceived as always under construction, in the process of forever being made, implying a continual openness to the future. These steps are fundamental to an appreciation of the deeply political nature of geography: "Conceptualising space as open, multiple and relational, unfinished and always becoming, is a prerequisite for history to be open and thus a prerequisite too, for the possibility of politics" (p. 59).

Finally, poststructuralism has also been accompanied by a questioning of the ocularcentrism so taken-for-granted by modernist thought. Vision, as Jay (1993) argues, is not simply a function of biology, but also an historically specific way of knowing the world. To visualize – to gain insight, to keep an eye on something – is to invoke a host of cultural and linguistic tools to make sense of reality. The equation of vision with truth, in this perspective, is essentially a positivist assumption that denies the possibility of other ways of understanding the world, particularly aural ones (Sui 2000). Rather than "looking down" on subjects and spaces, post-positivist modes of understanding include the tactile spaces of the body, the phenomenology of everyday life, and the centrality of language in the constructing of knowledge. Thus, poststructuralism has emphasized uncertainty and difference rather than certainty and similarity, truth as embodied, negotiated and produced rather than given. In place of the all-knowing Cartesian observer, poststructuralist thought has emphasized the positionality of the situated observer, whose viewpoint is always partial, incomplete, and power-laden. As Rorty (1981) famously noted, once we abandon the positivist metaphor of the mirror as the basis of objective knowledge, we are led to the metaphor of the conversation, in which language, positionality, and dialogue are central.

5 Geography, post-communism, and comparative politics

Jeffrey Kopstein

The waves of democratization and de-democratization that swept over the Eurasian landmass in the 35 years after 1974 spawned an industry of interpretation. In particular, the collapse of communist governments from Berlin to Ulan Bator after 1989 and the emergence of 28 post-communist states offered social scientists an irresistible laboratory for testing some of their most cherished hypotheses about the nature of political change and conditions under which democracies thrive or collapse. Multiple countries emerging from a form of government that imposed unprecedented kinds of institutional, economic, and social standardization offered a comparable starting point. Even more appealing for cross-national research, by the mid-1990s the variation in outcomes was already easy to see. Considering the similarity of these countries when they began their post-communist journeys, the huge variation in regime-types that quickly emerged cried out for some sort of explanation. What accounts for this variation? Why did some countries have it easier than others? Why were some able to quickly consolidate democracies (even with very little previous democratic experience), why did others fail to move far from authoritarianism or slide back to authoritarianism after an abortive flirtation with democracy, and why did a large group end up as hybrid regimes, neither fully democratic nor completely authoritarian but something in between?

The tools and concepts of political geography seemed especially appropriate for addressing this nexus of questions. First, since the countries of the post-communist world were all geographically contiguous, they could conceivably be thought of as constituting one big "region." It stood to reason that what happened in one place would influence developments in another. Second, there was an older, if not always respectable, German and central European tradition of political geography in the borderlands of Europe, some of it invoking conservative geopolitical thought to account for shifting fortunes of nations and empires. Other strands of continental thought attempted to explain the neat geographical variation in economic and political conditions already apparent to eighteenth- and nineteenth-century observers of the European continent with the notion of a *Kulturgefälle* or cultural gradient (Burleigh 1988). As questionable as some of these methods and concepts were, they did share an elective affinity with the intuitions of most Western experts on the region who, when asked in 1989 which countries would ultimately "succeed" (a concept best left vague), normally pointed to the states bordering Western Europe, with the variance in success and failure increasing as one moved further east and south across the continent.

Notwithstanding this conventional wisdom, a wisdom easily accessible to scholars and the general public on the pages of the *New York Review of Books* with article after article during the 1980s both calling for and describing the "return to Europe" of Hungary, Poland, and Czechoslovakia, this was not the direction taken by the subfield of specialists in political science in comparative politics in its initial accounting for change. The field of comparative politics, which had emerged after World War II in the USA, had never fully integrated political geography into its intellectual apparatus for the simple reason that it was set up to study the politics of "foreign countries" (Janos 1986). The essence of comparative politics is the comparative method and for political scientists the natural units of comparison are countries. Comparativists (as students of comparative politics refer to themselves) therefore tended to treat countries as discrete units, explaining the variation between them as the product of variations in what was going on within them individually. This method provided maximum inferential leverage by multiplying the number of cases with variation on both independent and dependent variables. In addition, for a generation of modernization theorists it opened the tantalizing possibility of some sort of universal history through which all countries pass in one way or another on their way to democracy. This orthodoxy was attacked by neo-Marxists and world system theorists in the 1970s. They pointed to the interdependence of cases – the North became rich and democratic, so the argument ran, because the South remained poor and authoritarian. This attack, however, lost traction when confronted with the success of Japan, Korea, and ultimately China in changing their position in the global hierarchy, which in turn was explained by the internal policies and features of these countries (Johnson 1982).

The first wave of comparativists therefore largely ignored the conventional wisdom and proceeded to explain variation in political and economic outcomes in the post-communist world by factors internal to the cases themselves. Hungary and Poland succeeded in consolidating their democracies because of features internal to Hungary and Poland. Kazakhstan and Turkmenistan remained authoritarian for reasons to be found within the economies and societies of those countries. Success and failure were explained as a function of institutional design, sequences of reform, and the ability or inability to marginalize opponents of democracy. The spatial turn in comparative politics came as a response to this first generation of explanations, which themselves seemed unable to account for why some countries chose the "right" institutions, why others were able to sequence their reforms properly, and why the enemies of democracy could be cast aside in some countries but not in others. In what follows, I outline this shift in more detail. Yet, the story does not end there. The next section shows that once the spatial turn had been accepted as identifying an important correlative relationship, once "geography" seemed to account for so much of the variation in post-communist outcomes, comparativists began to question this explanation and ask, what is the *real* variable for which geography is the proxy? Some maintained that the impressive statistical relationships were really tapping into deeper historical as opposed to spatial structures. Others contended that the real causal variable underlying the spatial dependence of outcomes was policy choices made in Western capitals, especially in Brussels. In the final section, I argue that these are important addendums to the spatial turn but they do not invalidate it. Even once other, competing explanations for the variation in post-communist outcomes are taken into account, there is an important residual element of geography and spatial dependence that cannot be dismissed or reduced to other factors.

From transitology to comparative politics

It should not be surprising that students of post-communist politics looked to comparative politics, which had been studying transitions to democracy since the beginning of the "third wave" in 1974. Of course, there was much debate, some of it heated, about whether the ideas and concepts of "transitology," which had developed using the cases of Southern Europe and Latin America, were appropriate for studying the post-communist experience (Schmitter and Karl 1994; Bunce 1994). The critics of transitology not only remarked on its teleological character – it appeared to anticipate no other possible outcome than democracy – but also its lack of attention to the history of particular cases. It was probably inevitable, however, that once the totalitarian regimes of Eastern Europe disappeared, students of other regions of the world would find it easy and irresistible to apply their methods and concepts to a new region. Additionally, the kinds of questions that the transitologists asked *were* important ones. What are the modal sequences by which authoritarians cede power to those committed to multiparty elections? Under what conditions is the transition peaceful or violent? (Linz and Stepan 1996; O'Donnell and Schmitter 1986). How can a sense of community be reconstructed after a brutal dictatorship? How can the competing demands of different ethnic communities be accommodated? Do some kinds of constitutional structures and political institutions work better than others? Is it wiser for post-communist leaders to privatize and marketize their state-run economies quickly or does this cause political instability?

The transitologists never claimed that democracy was inevitable, but the answers to these questions implied that whether it did take root was a function of human will and choice. The literature on transitions to, and consolidation of, democracy expressed a deep commitment to the importance of human agency (DiPalma 1990). In doing so, it was responding to an earlier generation of theorists who claimed to have found a set of preconditions for democracy, the most important one being economic development (Lipset 1963). Yet the collapse of communism and the rapid fielding of multiparty elections almost everywhere appeared to demonstrate that there were no preconditions for democratic rule, or, if such preconditions existed, they were minimal and could easily be compensated for by committed leaders, sensitively handled transfers of power, and cleverly crafted institutions.

Comparativists took the next step and attempted to identify the conditions under which democracy became "the only game in town" (Przeworski 1991). Using cross-national research designs and drawing on the experience of Latin America and Southern Europe, students examined the impact of different executive-legislative and electoral system designs on democratic outcomes. One finding was that, the stronger the presidency, the less likely a country was to become and remain a democracy (Linz and Stepan 1996; Fish 1998). The lesson was clearly to choose the right institutions. Other comparativists remarked on the importance of driving the communists from office quickly in order to set the stage for good economic policy and economic recovery, which in turn would help consolidate democracy (Fish 1999).

It is appropriate at this point to note that some comparativists maintained that while comparisons are crucial, the concepts and ideas drawn from other regional contexts might not easily travel to the post-communist world. The most salient distinguishing feature of the post-communist context, of course, was the experience of communism itself (Jowitt 1992; Hanson 1995). Empirical research quickly confirmed that there was indeed something different about the communist world. For example, Howard's (2003) cross-national research on civil society showed membership in social organizations to be systematically lower in all post-communist societies than in other formerly authoritarian countries. Likewise, transition economics repeatedly noted just how different the political economy of post-communism would be due to the lack of a pre-existing moneyed middle class. Last but not least, the Balkan wars of the 1990s and at first hot and then frozen conflicts in the southern portions of the former Soviet Union demonstrated how difficult it would be to construct viable national communities after decades of suppressing ethnic identities.

But if the legacies of communism were ubiquitous, their impacts were unevenly distributed. Some countries managed to establish stable institutions of democratic representation, viable market economies, and reasonable modes of intercommunal relations. Others could not. If the difference between the cases of success and failure was really one of human will that set some countries on the right path and others on the wrong path, what explained this distribution of choices? In fact, looking at the map of the formerly communist world, it became apparent that the virtues associated with the right choices (parliamentary versus presidential government, electing the communists out of office in the first election, quickly marketizing the economy, finding a mode of coexistence between ethnic groups) were distributed in a remarkably neat and regressive geographical pattern across the Eurasian landmass. If post-communist outcomes were path-dependent, if there was a significant lock-in effect from the initial institutional and policy choices made by post-communist elites, then a natural question to ask was, what determined the path?

The spatial turn

Drawing on important works within economic geography, some economists pointed to the spatial advantages and increasing returns to scale associated with access to markets in the West enjoyed by countries bordering on the European Union. The spatial revolution quickly spread beyond economics, however. Widely available GIS software and computing capacity permitted the identification of a broad range of patterns in social and even political developments in the post-communist world. Furthermore, these spatial patterns could be tested against both accounts, i.e., those emphasizing policy and institutional choices and those underlining the impacts of communist legacies.

Kopstein and Reilly (2000, 2003), inspired by the work of political geographer John O'Loughlin and assisted by the statistical software of Luc Anselin (O'Loughlin *et al.* 1994), showed that the evidence for the spatial determinants of post-communist outcomes was just as compelling as the evidence for temporal path-dependence. They suggested that the diffusion of the resources, institutions, and norms necessary to post-communist success had been proceeding in a geographical pattern across the Eurasian landmass. Several tests of this proposition were performed, each designed to get at an element of the phenomenon. The crudest test was the insertion of a control variable for "distance from Berlin or Vienna, whichever is closer" from each post-communist capital city into pooled cross-sectional time series, using yearly data, with the dependent variable being Polity IV scores for democracy minus autocracy (a standard measure used in political science). The main control variables deployed measured whether or not the communists were removed from office the first elections and the Dow Jones score for bureaucratic rectitude for each post-communist country. The results were highly suggestive and demonstrated that spatial explanations, even crude ones, could hold their own against two of the most cherished hypotheses of the comparative politics literature – one stressing the impact of the transition itself and the other focusing in on the long-term effects of different legacies of bureaucratic efficiency.

A more sophisticated test using scores of each post-communist country's geographically contiguous neighbors to predict the Polity IV democracy scores yielded even more powerful statistical relationships, suggesting that not only a country's distance from the West but its location, no matter where it may be, exercised a profound impact on the post-communist experience. This neighborhood effect helped countries if their neighbors were democratic and hurt countries if they were caught in a zone of authoritarianism, war, or religious fanaticism. Furthermore both the Gi* and the Moran's I statistic showed a high degree of spatial autocorrelation, something best visualized on maps (which are still rare in political science journals). Certain countries, such as Germany, Austria, Afghanistan, and Uzbekistan, appeared to exercise especially powerful effects on their neighbors, suggesting that with more effort and better data these spatial relationships could be teased out with greater specificity.

In addition to these statistical tests, case studies also showed the powerful influence of neighborhood. Two interesting cases here were Slovakia and Kyrgyzstan. After communism, Slovakia systematically pursued what political scientists characterize as "bad" policies. Under the would-be authoritarian Vladimir Mečiar, property was privatized to cronies, political opponents were harassed and jailed, and the Hungarian minority's language rights were continuously put under threat. Mečiar initially succeeded in dividing the opposition, who spent more time denouncing each other than in organizing against the creeping dictatorship. Yet, the threat of exclusion from both NATO and the European Union at a time when the country's two primary external referents, Hungary and the Czech Republic, were about the gain entry to both, encouraged the opposition to overcome its own internal problems and unite to defeat Mečiar in a national election. Research also showed that the opposition benefited from the attention lavished upon it by democratic parties and governments in neighboring countries, such as Austria and, slightly more distant, Germany, but also from its partners in the Visegrad group. In Slovakia's case, good geography helped overcome bad policies and choices. The same goes for Hungary, Croatia, and Poland, all of which were able to suppress the most pernicious forms of nationalism in the hopes of gaining entry to the European Union.

No such logic holds for Kyrgyzstan, where bad geography ultimately trumped multiple efforts at good policies. When the Soviet Union broke up in 1991, Kyrgyzstan was led by a former physicist, Askar Akaev, who had a reputation for being liberal and committed to integration with the West. His country quickly left the ruble zone and was the first post-Soviet state to join the World Trade Organization. Notwithstanding these achievements, the country's democracy ratings steadily deteriorated. For one thing, Western attention was sporadic. Foreign investment tended to be purely resource-based and frequently sought licenses and advantages by corrupting the government. Kyrgyzstan's powerful neighbors, especially Uzbekistan and Kazakhstan, consistently applied pressure on democrats with the intention of heading off any contagion into their own dictatorial states. Akaev's rule devolved into a dictatorship which was ultimately overthrown in the wake of a rigged election in 2005. Yet, despite the attempt at democratic renewal, Akaev's successor, Kurmanbek Bakiyev, found equally few consistent friends in the West and quickly began subverting the democratic gains of the "tulip revolution." The most interested parties remained the local and powerful neighboring dictators in Central Asia and Russia. The result has been a steady slide back toward authoritarianism.

Lankina and Getachew (2006) extend the spatial diffusion research program in an interesting way. Whereas Kopstein and Reilly found no evidence for the impact of diffusion from the West on Russia's politics (hypothesizing that the large "civilizational" countries may be more impervious to outside influences), Lankina and Getachew demonstrate that spatial diffusion appears to be at work not only between countries of the post-communist world but also within them. They attempt to explain variation in levels of openness and democracy in Russia's 89 regions. Even with authoritarian leaders dominating the national stage, they find that "diffusion processes and targeted foreign aid help advance democratization at the sub national level in post communist states and other settings." Their study makes the case for a "geographic incrementalist" theory of democratization through a statistical analysis of over 1,000 projects carried out by the EU in Russia's regions over a 14-year period. The EU's commitment to democracy is especially apparent in regions located on its eastern frontier. This interest, as well as other processes of diffusion from the West, has systematically made these regions more democratic over time even if they began their post-communist journey more closed than their eastern regional counterparts.

The use of concepts and methods from geography draws our attention not only to important sources of political change, it also helps us get a purchase on our emerging mental map of the globe's regions. As the processes of change unfold, political geography helps us understand how new regions take shape out of a formerly singular post-communist space. As these spaces become invested with meaning (a process in which scholars may influence the reality they analyze), they will make the transition from spaces to places.

Lankina's and Getachew's observation, however, that the EU provided the push for this redefinition of post-communist spaces raises the crucial question of whether geography is simply a proxy for other causal processes. If true, then perhaps "space" and "place" are not the genuine sources of variation in outcomes. Perhaps other variables are "doing the work." This idea, in fact, has been the dominant reception of the spatial turn within comparative politics. The reaction of most comparativists, rather than seeking to advance the spatial diffusion paradigm within the discipline, has been to look for more powerful domestic temporal processes that have shaped politics in the region.

Although the correlations between space and location, on the one hand, and political outcomes, on the other hand, are powerful, what exactly is performing the heavy lifting in geographic explanations? In the Slovak case, it was the prospect for joining Western economic and security structures that created the impetus for democratic change. Geography, then, seems to be a reasonable proxy for the geopolitical design of Western international institutional builders who desired Slovak membership in order to shore up Western Europe's eastern periphery. In the Kyrgyz case, would-be democrats had no such institution to which to refer. On the contrary, those pointing to friends in the region could only identify autocrats.

These logics, however, have not been entirely convincing to students of comparative politics. Instead, comparativists have sought greater clarity in causal chains linking space or place to political outcomes than has been provided by scholars to date. It is to these studies that we now turn.

Geography as a proxy: for what?

Even when rejecting the use of concepts and methods from geography, a number of authors working on post-communism have attempted to address the spatial turn. Judging by its frequent inclusion in multivariate models of the determinants of democracy and dictatorship in the post-communist region, political geography has been accepted within the subfield. The consensus, however, is that the few extant efforts to use the insights of political geography to account for variation in regime-type outcomes have been more suggestive than decisive. Comparativists have generally argued that the statistical evidence for spatial dependence is really tapping into other, deeper factors besides geography.

Kitschelt (2003), in particular, maintains that the neat geographically regressive pattern of development of the post-communist world is really a function of pre-existing patterns of modernization. The notion that the starting point for the post-communist countries was essentially the same, he argues, is mistaken. Instead Kitschelt identifies three different kinds of communism, each corresponding to different state traditions. "Bureaucratic communism," relying on a pre-existing state with high levels of bureaucratic rectitude, dominated East Germany and Czechoslovakia. "National accommodative communism," which attempted to reconcile communist rule to patterns of a highly mobilized civil society, was the model in Hungary and Poland. The third model, "patrimonial communism," established itself in Romania, Bulgaria, and presumably throughout the non-Baltic Soviet Union, where the communists, upon seizing power, confronted pre-bureaucratic state structures. Patterns of communist state-building in turn shaped both patterns of civil development and bureaucratic efficiency passed on to the post-communist democracies. In short, communism was forced to build states and manage societies with the raw material at hand and these states and societies bequeathed to the post-communist world much of the same bureaucratic and civic cultures of the pre-communist era. It is these kinds of causal chains that remain most appealing to students of comparative politics.

Similarly, Darden and Grzymała-Busse (2006) also maintain that geography is primarily tapping into the modernization phenomenon and the determinants of both the mode of exit from communism and post-communist political change are more readily found in the formative nation-building experiences of particular political units across the continent. If a group became literate before communism, they maintain, it was far more likely to cast off the communist legacy and consolidate democracy after 1989 than if it did not. Again, for Darden and Grzymała-Busse the long-term historical causal chains are the point where the mind rests. The key to a good explanation for most comparativists, then, is to look for critical historical junctures within societies rather than identify contemporary or historical confining conditions in processes that occur between them.

The most extensive statistical test to date of this debate has been performed by Pop-Eleches (2007). For the most part, he concurs with Kitschelt and Darden/Grzymała-Busse. Numerous model specifications and conceptualizations of dependent variables for democratic and non-democratic outcomes (Polity IV, Freedom House, and Voice and Accountability), show that historical legacies continue to predict democracy and autocracy well even when spatial factors are taken into account. Equally, however, geography cannot easily be eliminated as an independent explanatory factor in accounting for post-communist political outcomes.

If some comparativists maintain that geography is a proxy for historical legacies, others have argued that it is mostly an expression of the politics of EU enlargement. The strict regime of conditionality and monitoring imposed on the post-communist candidate states and the requirement that they pass the entire corpus of EU law into national legislation decisively influenced almost every aspect of politics in the region. In everything from minorities policy to the holding of free and fair elections, the prospect of membership determined the course of events through 2008. Geography's statistical significance and substantive effect, so the argument runs, is actually an artifact of the EU's decision to admit countries geographically contiguous with existing member states. Of course, this argument in no way contradicts the notion that geography drives events. It simply pinpoints geopolitics and trade interests as the underlying force for why location matters.

Vachudova (2005) distinguishes between the EU's passive versus active leverage in post-communist Europe. Immediately after 1989, the prospect for admission did alter preferences in the region but not, Vachudova contends, in decisive ways. It was not until the decision was made in 1999 to begin negotiations that that the leverage of the West over the East moved from being passive to active and the regime of conditionality and monitoring really began to shape events. Cameron (2007) puts the date of active leverage much earlier, to the early 1990s, when the post-communist states began to sign Association Agreements with Brussels. These agreements in essence amounted to "holding pens" for countries while the member states decided upon whether and when to begin formal negotiation on membership. Both Vachudova and Cameron are correct in this respect. The EU has much more direct influence over Croatia and Turkey, as both actively negotiate for membership, but nevertheless retains an important voice even within Serbia, which has expressed interest in signing its own Association Agreement. Both authors, however, pitch their arguments in opposition to spatial explanations. Cameron even deploys spatial control variables in his main regression.

Theoretically the EU could decide to admit countries that do not share borders at all with the EU and membership conditionality would still profoundly shape political outcomes. In this sense, geographic diffusion and EU conditionality are conceptually distinct phenomena. The fact remains, however, that the EU's decisions on where it will enlarge and which countries start down the road of negotiation are deeply political and this political logic is intimately related to geopolitics and trade policy. The discourse on Europe's borders and whether or not Turkey's negotiations should be drawn out or cancelled reflects deep divides over whether it is a good idea for a poor and non-European (read: Muslim) country to join the EU. In short, it is difficult to disentangle the logic of EU conditionality from the geopolitics of enlargement.

In what is probably the most comprehensive treatment of the relationship of EU conditionality to the role of geography, Levitsky and Way (forthcoming) conceptualize the matter as one of "linkage" and "leverage." Linkage refers to the connections between countries that may occur through trade, investment, tourism, and the like. Leverage, on the other hand, refers to the power or influence of international organizations over a target country that desires membership, approval, or resources from the organization. Although the authors acknowledge the leverage exercised by the EU over candidate states, they also argue that in the 1990s geographical proximity to Western Europe promoted important linkages that weakened autocrats even in countries with little immediate prospect for membership, such as Macedonia, Albania, Croatia, and Serbia. "[I]n each of these cases, Western intervention significantly weakened autocratic incumbents and strengthened democratic forces – leading to democratization in Croatia and Serbia and near democratization in Albania and Macedonia by 2005" (Levitsky and Way, forthcoming). This line of argumentation suggests that rather than geography being a proxy variable for EU membership, the reverse may in fact also be true: EU membership is actually a proxy variable for geography. Even without the straightjacket of membership conditionality, geography may influence outcomes by way of linkage rather than leverage.

In fact, it may be the case that some version of "pure" geography may influence political outcomes independent of any linkage or leverage. The evidence from post-communist Europe suggests that people living in proximity to the West believe that their fate somehow matters more to Europe and Europeans than those further East. They think of themselves as Europeans and consider "European standards" as something worth respecting. Of course, this sense of Europeaness is historically constructed and reconstructed, yet the fact remains that it is easier to make European arguments in places closer to the entity that now calls itself Europe – the EU – than in locations that are far way and about which Europe cares little. This geography of affection and knowledge is difficult to measure but there can be little doubt that it has helped shape the post-communist political landscape.

Geography, post-communism, and comparative politics

The subfield of comparative politics may represent a hard case for sustaining the influence of spatial concepts and methods. What accounts for the reluctance of students of comparative politics and post-communism to advance the spatial paradigm rather than attempting to refute it? It is certainly possible that it is a purely a matter of evidence. If the evidence sustained the assertion that spatial diffusion determined outcomes more consistently than did internal politics, political geography would be accepted as part of the mainstream of the field. Political geography remains marginal, so the argument runs, because it deserves to stay marginal.

The evidence, however, that evidence matters in the development of the social sciences remains meager at best. Paradigms neither rise nor fall because of evidence. A much more likely candidate for explaining the continued marginalization of political geography within comparative politics is to be found in the specific history of comparative politics within North American political science. The field of political science, as it developed in the United States after World War II, divided into multiple subdiciplines. One of the most important disciplinary divides was that between international relations, which studies primarily war and trade between states, and comparative politics, which was originally set up to study the domestic politics of "foreign countries." This division of labor made a certain degree of sense, since the latter pursuit required deep immersion into exotic histories and frequently difficult foreign languages. Yet, it was probably equally silly to believe that patterns of war and trade could be explained without reference to domestic politics as it was to believe that the domestic politics of states unfolded as if they were not located in given locations, with particular neighbors, and specific physical geographies.

Of course, students of both subdisciplines quickly recognized this, but there quite understandably remained a disciplinary bias within each field against accepting methods that question its assumptions. All disciplines have conceits and the major conceit of comparative politics is that, at a minimum, mid-level generalizations and patterns can be gleaned by comparing the politics of two more states. Simple enough and probably true. But the comparative method depends on one basic assumption: unit homogeneity. If the politics of the units under observation are influencing each other, then the assumption is violated, the "experiment" is contaminated, no generalizations or patterns can be found, and the entire field of comparative politics is called into question. The methods of political geography, especially those involving statistics measuring spatial autocorrelation, were designed almost specifically to question the assumption of the independence of the units under observation, for the simple reason that they measure just how much the units do influence each other. Should we be surprised at the resistance to methods and concepts as mainstream modes of analysis when these methods and concepts threaten the very foundations of the field? This issue, of course, is more a matter of the sociology of inquiry than of inquiry itself but it probably better accounts for the place of geography in comparative politics than any adjudication of the evidence or model of normal science.

Even with this resistance, political geography refuses to leave comparative politics alone. Discussions of Galton's problem (which posits the difficulty in using cross-cultural data because of external dependencies), "regions" in world politics, and the role of "public space" in channeling political discourse represent but a few of myriad ways in which spatial thinking has infected the subfield. One suspects at some point that testing for spatial effects in cross-national research will become standard in the best journals, if for no other reason than it will help decipher some of the important puzzles in comparative politics.

6 Retheorizing global space in sociology

Towards a new kind of discipline

Harry F. Dahms

As one of the blind spots in sociology, the neglect of *space* is both prominent and instructive. While *time* has played a comparatively central role during the disciplinary history of sociology, space did not play as explicit and visible a role, nor did the category seem to be relevant for sociological research (Roy and Ahmed 2001). Indeed, even a cursory review of the sociological literature, especially during the decades following World War II, reveals research interests and practices that did not provide much room for space and place – as systematic concepts or legitimate research concerns.[1] The area in sociology in which space played the most important role was urban sociology, whose origins date back to the 1920s and 1930s, and which experienced a rapid expansion during the 1970s and 1980s.[2] Even though – or possibly, because – all sociologically relevant phenomena occur in space and time, for the vast majority of sociologists, space may have been too obvious a feature of human existence to warrant careful attention with regard to its potentially specific features and impact on social life. Since the implicitly presumed purpose of both theoretical and empirical research in sociology was the production of knowledge about conditions in all "civilized" and complex societies independently of, or rather, beyond space and time, concrete features that influence social forms in clearly delimited contexts were regarded as secondary (Dahms 2008). As a consequence, the history of sociology has been burdened by an odd paradox.

On the one hand, the idea and especially the discipline of sociology could only emerge in the particular historical context that resulted from the continuous reconfiguration of social, political, and economic life necessitated by the emergence of a market economy during the late eighteenth and early nineteenth centuries in Western Europe and North America, including the formation of nation-state, the onset of the industrial revolution, and the emergence of the modern labor movement. On the other hand, sociologists have been dedicated to identifying general principles and "laws" of motion in society that were not meant to be limited to modern society alone. Even though concrete socio-historical contexts served as the foil for arriving at insights about characteristic features of the genus *modern society*, most sociologists during the twentieth century, especially after World War II, applied a mind-set that did not distinguish explicitly between knowledge about society in general and knowledge about modern society as a specific form of social organization. Rather than acknowledging that sociology by default is the social science of modern-industrial-capitalist-democratic society, most sociologists were not willing to be sufficiently reflexive regarding the discipline's specific spatial and temporal conditions of emergence to achieve clarity in this regard.

For instance, many sociologists recognized and studied the sociologically relevant patterns that applied in more or less similar ways in all societies that had spawned sociology: France, Britain, Germany, and the United States, above all. Yet at the same time, they regarded as the main purpose of the discipline the production of knowledge relating to those patterns as relevant beyond the limitations of their contexts of origin. In order to put the resulting knowledge to good use, for the betterment of conditions that ranged from concrete circumstances in particular societies, to "humankind," sociology presumably had to be oriented toward an understanding of the nature of society in general, not just modern society. In retrospect, what arose as modernization theory during the 1950s was driven by the conviction that there is a logic at work in all societies that, once understood and appreciated, would enable decision-makers and collective actors to work to insure that all societies on Earth would follow a predictable track toward prosperity, security and peace that was modeled after the modern societies of Western Europe and North America.

For the most part, however, the mind-set that produced modernization theory, along with closely related approaches, impeded consideration of the specificity and relative autonomy of socio-historical contexts, of space and time as sociologically central categories that necessitate careful examination of how and to what degree what appeared as more or less disembodied patterns, principles or laws in fact were tied to the particularity of specific environments. Put differently, the history of mainstream sociology represents a set of stories about how social research was inspired by the notion that the specificity of socio-historical contexts was ancillary to general principles that shaped, and continue to shape, the concrete form of those contexts.[3] Sociologists who were critical of mainstream approaches, on the other hand, contended that the discipline is both uniquely conducive to and responsible for devising and applying frameworks that enable social scientists to recognize the dynamic entwinement of general principles and the socio-historical specificity of concrete forms of social life. Indeed, the effort to differentiate clearly between general principles and concrete contexts impedes appreciation precisely of the kind of dynamic processes that made (and continue to make) possible modern society as an actually existing form of social organization.

Two questions regarding the initial neglect of space in sociology are especially important. First, was this neglect a consequence of how the discipline's classics framed the systematic study of society, with regard to the agenda, responsibility, and tools of necessary and desirable social research?[4] Second, is the neglect of concrete environments, as they may influence, shape, and reinforce social relations, forms of interaction, and modes of coexistence, indicative of a larger (yet latent) problem in the practice and orientation of sociology that must be spelled out explicitly, as far as taking into consideration the multiplicity of dimensions of societal reality is concerned, including space?

Space in classical social theory

As far as the classics are concerned – especially Marx, Durkheim, Weber, and Simmel, but also Charlotte Perkins Gilman and W. E. B. DuBois – we must distinguish between explicit consideration of space and spatial categories, and the role of space in the framing of their theories. As John Urry (2001: 4–5) pointed out, there is a striking absence in classical theories of references both to space and to categories that are sensitive to the potential linkages between space and social forms – with one exception. Georg Simmel conceptualized space as entailing meaning, in several ways: space is exclusive or unique; divided and "framed"; the site of social interactions; a determining factor for proximity between individuals (or lack thereof); and the opportunity for geographical mobility.

To be sure, all the classical theories were compatible with recognizing the importance of space, and it should have been possible for sociologists in the twentieth century, and especially after World War II, to both recognize, and seize upon, that compatibility. For instance, in one obvious regard, the theories of Marx, Durkheim and Weber directly related to geographical as well as organizational space: in different ways, each was concerned with the fact that modern society (qua bourgeois society, capitalist economy, and nation-state) emerged in Western Europe, as a very specific and clearly delimited geohistorical space.[5] Implicitly, explicitly, or by default, they also were comparative social scientists: Marx lived in different countries, and he could not avoid being aware of the fact that social circumstances differed from place to place, from city to city, from country to country; Weber engaged in an expansive comparison of geographically based world religions; Durkheim grounded sociology in the distinction between premodern and modern societies, and relied on comparisons between modern societies (e.g., between France and Germany, in his study of suicide). In addition, these classics were concerned with ongoing organizational changes as a central feature of modern society: Marx regarding the site of productive labor (from farms to mines and factories); Durkheim in terms of the accelerating differentiation of human labor in society, as manifest in institutions and organizations; Weber with regard to changes in organization, the emergence and transformation of business enterprise, the unstoppable spread of bureaucracy, and especially the sociological importance of the city (Isin 2006; Turner 1999: 214–5).

We should keep in mind, though, that since efforts of the classics in different ways were directed at engendering and grounding rigorous analyses of the nature of social life in the modern age, it would not have been possible for them to address *all relevant dimensions* of social life, even if they would have tried to do so.[6] Thus, with an eye towards the history of sociology during the twentieth century, the important question is not so much whether the classics did justice to space, but whether, and to what degree, their theoretical designs were compatible with and conducive to consideration of the spatial dimension of social life. In an assessment of the influence of the discipline's founders on the ability and willingness of sociologists today to recognize space as a key category, however, we need to keep in mind the likelihood of there being a discrepancy between how each of the classics conceived of the promise and responsibility of sociology, and how later sociologists interpreted and applied the earlier frameworks. Since such a discrepancy is rather inevitable, the main issue is its nature and depth. The challenge regarding assessments of how the classics dealt with space and place, then, pertains to the question of how to identify and interpret the efforts of the classics in the present context, and how adequate prevailing interpretations of their theories during the twentieth century were in light of recent reassessments.

The challenge of theorizing modern society

With globalization providing one of the most important research challenges for social scientists today, certainties about well-established ways of reading Marx, Durkheim, Weber, Simmel and others, have begun to give way to reassessments of the current value of their contributions. If the kinds of reading and interpretation that informed and oriented the disciplinary history of sociology during the twentieth century were based on flawed appropriation of the classics' contributions, conceptions of the purpose of sociology since the classics should have been detrimental to realizing the discipline's promise.[7] In retrospect, it would appear that, especially after World War II, in the context of the Cold War, the majority of sociologists were less inclined to accept and apply the radical thrust underlying each and all of the classical theories, as imperative to the design and execution of social research.[8] In his most recent book, Charles Lemert (2007) described the continuing relevance of the classical theories as projects of "thinking the unthinkable." By contrast, most sociologists during the second half of the twentieth century did not read the classics as having been engaged in efforts to frame sociology in ways that require the willingness and ability of sociologists to "bend" perspectives and mind-set regarding societal reality we all acquire through socialization, education, and social interaction. In different regards, each of the classical theorists conveyed the importance for sociologists to make the following distinction. In one regard, as members of society, we are both compelled and inclined to subscribe to perspectives on social life that are a precondition both for our ability to "function" in society as individuals, and for the possibility of a stable social order. In another regard, for sociology as a social science to be an actual possibility, as sociologists we must be critically and rigorously reflexive regarding precisely those perspectives on social life as potential impediments to our ability to confront the actuality and underlying logic of societal reality. Paradoxically, however, after World War II, sociologists tended to assume that the victory of democracy over "totalitarianism" constituted a major break in the history of human societies. From that point on, in the spirit of the Enlightenment, modern Western industrialized societies would socialize and educate individuals to a growing extent in ways that were conducive to recognizing the reality of social life and one's own place within it. Consequently, it would no longer be required of social scientists to make a determined effort to dissect preconceived and socially condoned notions about reality with the same rigor and to the same degree – as the necessary precondition for scrutinizing modern society, as each of the classical theorists had emphasized with regard to a particular dimension of social reality – in Lemert's (2007) language, as "thinking the unthinkable."

For the present purpose, to illustrate the gap between predominant readings of the classics during the post-World War II era, and new ways of reading that have become possible under conditions of globalization, suffice it to suggest one alternative mode of reading the classical contributions. If we start out from the centrality to their respective concerns of the concepts of alienation (Marx), anomie (Durkheim), and Protestant ethic (Weber), and rather than reading the three theorists against each other, read them together, the following would be one possible scenario regarding the constitutional logic of modern society.

While economics started out as a discipline concerned primarily and affirmatively with the necessary preconditions for the successful pursuit of prosperity, both at the individual and the national level, sociology emerged as the discipline that was concerned with the "price" society must pay, in order to enter and remain on the track of increasing prosperity.[9] In different ways, Marx, Durkheim, and Weber were concerned primarily with the kind of continuous societal transformations necessitated by the direct and indirect orientation of political, social, and especially legal institutions toward providing and maintaining an environment that is conducive to profit-making – i.e., a social order characterized by a high degree of stability. Within this interpretive framework, Marx was primarily concerned with the spread of the capitalist mode of production engendering processes of social, political, and cultural adaptation and organization that carry a destructive potential. In addition, they bring about qualitative changes in the nature of the relationship between individual and reality at all levels, especially regarding one's own self, others, nature, and the "species." Durkheim was concerned with the consequences resulting from the continuous differentiation of labor in society, for the nature of social relations, and the combined danger of individuals failing to maintain their place in the social environment, as values lost their strength, and the moral fabric of society eroded. Weber, finally, was concerned with the question of why modern capitalism only emerged in the West, and how the reconstruction of Christianity as a consequence of the Reformation went hand in hand with economic advances and transformations that produced an entirely different societal reality, compared to the Middle Ages, triggering economically driven processes of rationalization in all spheres of life that were increasingly detrimental to the meaningful interpretation of human existence.

If we contrast these three theories (representing the plurality of classical theories), as was common during much of the twentieth century, the primary question will be which of the theorists "was right," and which were not – and to what degree and in what regards. However, if we read these theories as complementary, a different reading ensues that is highly compatible with, and indeed conducive to, considering the importance of space in today's world. In fact, recent attempts to stress and illustrate the importance of space thus appear as efforts to identify, as the charge of sociology, a type of inquiry that is characterized by continuity with the classics' agendas and conceptions of sociology as a social science. The challenge, then, is to pinpoint those dimensions of social reality whose neglect would be detrimental to the possibility of systematic research relating to contemporary social life, and to endeavor to construct a theoretical foundation for social research that combines those dimensions. Space most certainly would have to be one of those dimensions.

A synchronized reading of Marx, Durkheim, and Weber would be conducive, for instance, to perspectives on modern society as a force-field whose primary imperative is the stabilization of social order in a context that perpetually produces a socially destructive potential – *with both the destructive potential and the stabilization of social order being constitutional features of modern society*. Marx described the destructive potential in terms of alienation and commodity fetishism (followed in the writings of later theorists in the tradition he initiated, as reification, instrumental reason, and functionalist reason). By contrast, Durkheim was most concerned with anomie, and Weber with loss of meaning in what used to be framed in terms of the "iron cage."[10] In all of their theories, modern society appears as a sort of "container" that must be maintained, reinforced, and protected, for the continuously growing, economically beneficial, albeit socially destructive potential that accompanies capital accumulation and technological development to remain under control. The destructive potential, however, is an integral precondition for and consequence of a force-field that is in constant danger of collapse, and yet appears to be characterized by an unprecedented stability, as the consequence of an exchange of energy that must be carefully maintained. This stability is unprecedented both with regard to degree and kind of stability. On the one hand, there is a dynamic economic system that provides the energy that generates change in modern society, and that produces the resources that make modern society possible. On the other hand, politics, culture, legal system, and civil society adapt to increasingly dynamic economies in ways that provide a largely static shell. While economic processes constitute qualitative change, and are the cause of continuous political, legal, cultural, and social adaptation, the latter produce the *appearance* of change, distract from economic processes as the actual origin of change in the modern world, and conceal the absence, in reality, of qualitative change in politics, culture, legal system, and civil society.[11]

The main difference between this reading of the classical theorists and the mode that influenced and shaped the history of sociology is that the latter appropriation was inherently *static*, while the former is fundamentally *dynamic*.[12] In certain regards, it would be quite apt to interpret the efforts of poststructuralists, and postmodernism generally, as attempts to capture for the first time the kind of perspectives on modern society the classics were trying to engender, and which were far more intricate, as well as critical, than post-war sociologists were willing, or able, to recognize. In a sense, a certain strain of postmodernist, poststructuralist, and even postcolonial critiques of concepts of "the modern," then, would have been attempts to make up for the flawed appropriations of the classics in mainstream sociology. Rather than Marx having been a "structuralist," Durkheim having been a "determinist," and Weber having been a "positivist," their frameworks were directed to a far greater extent at facilitating a kind of critical understanding of social life that would have been comparable to the revolution that was occurring a century ago in theoretical physics.[13]

As an academic discipline, the history of sociology began at approximately the same time at which theoretical physics underwent a major revolution, in the form of Planck's quantum theory and Einstein's relativity theory. Yet most sociologists have worked with an image of theoretical physics that was tied to the universe of Copernicus and Newton. Very simply put, the difference between sociology as "inspired" by an image of physics modeled after Copernicus and Newton, and the kind of sociology that would have resulted from an orientation toward Planck and Einstein, is that the former is limited to the world of appearances, while the latter is driven by the determination to pursue the nature of reality as far and as deep as possible. My contention is that, even though disciplinary sociology emerged precisely at the time when modern physics underwent its most radical transformation, during the twentieth century, the vast majority of sociologists worked with an image of science that had become anachronistic at the beginning of the century – theoretically speaking. Furthermore, if more sociologists had been aware of developments in theoretical physics in the early twentieth century, they might have seized upon the opportunity to conceive of their discipline as having a systematic interest in distinguishing between surface appearances and underlying forces. Instead, mainstream sociologists appear to have dedicated their efforts at advancing sociology as a project that endeavors to analyze and *explain* surface appearances, without a rigorous concept of underlying forces.[14]

Sociological deficits and the challenge of globalization

As is true of all academic disciplines, the history of sociology can be told in terms of both its successes and its failures. As sociologists have labored to illuminate conditions of human existence in the modern world, they also have emphasized certain dimensions of social life at the expense of others. Succinctly put, the official history of sociology as a discipline is, for the most part, about how modern society manifests itself at the surface – at the level of discernable groups, institutions, organizations, and processes.[15] Despite a continuous proliferation of approaches that are critical of the orientation toward surface manifestations of societal life, as a mainstream discipline, sociology to date has refrained from taking responsibility for scrutinizing the underlying forces that produce the surface appearances. Instead, within the mainstream, to the extent that the link between both levels is being acknowledged and addressed at all, it tends to happen in passing, mostly in theoretically oriented projects and the related literature. In empirical research, the notion of underlying forces is almost entirely absent. As a result, a culture has taken hold in professional sociology that treats, as a matter of course, the analytical challenge as the explication of surface phenomena with reference to other surface phenomena – thus generating a kind of redundancy that is at odds both with the nature of modern society (as outlined above) and with current problems and challenges endemic to the context of "globalization."

Yet the efforts even of prominent sociologists who are critical of the discipline's core to engender a more reflexive disciplinary practice that would compel and enable sociologists to recognize illuminating globalization as a defining challenge of the present time (for instance, as the culmination of the contradictory trends that have been shaping the modern age[16]) so far have remained unsuccessful. Indeed, while the proliferation of approaches that point out problematic aspects of mainstream sociology may create the appearance of both the mainstream undergoing qualitative changes, the evidence is still inconclusive as to whether such indeed is occurring. The growing difficulties of representatives of mainstream approaches to deny types of social research that are critical of the mainstream a modicum of legitimacy are not necessarily, and certainly not automatically, indicative of the mainstream changing in profound and lasting ways.

For instance, rather than participating in efforts to reformulate and reconstruct the promise and responsibility of sociology in and for the early twenty-first century, mainstream sociologists tend to resist opportunities to tackle the gulf between conceptions of the discipline that regard the tensions between surface appearances and underlying forces as central, and conceptions that purport that such distinctions are immaterial to sociology. Concordantly, while mainstream conceptions tend to regard the potential practical and policy-related irrelevance of sociology as a non-issue – as its relevance is purported to be in evidence, and measured, in the number of academic departments and research institutes – critical conceptions are the expression of endeavors to insure the relevance of sociology in changing socio-historical contexts. Yet in light of difficulties to grasp the multifarious dimensions of "globalization," in particular, evidence keeps mounting that the proliferating contradictions of modern society are increasingly difficult both to deny and to illuminate in ways that translate into effective public policies in terms of stated goals, along with qualitatively superior perspectives on current conditions more generally (Dahms 2005). While the prevalence of contradictions can be observed without effort, given that their consequences are ubiquitous at the level of surface appearances – it is not possible to explain them at that level also.

Despite ongoing efforts, especially on the part of critical theorists, to stress and illustrate the need for sociologists to analyze tensions between surface appearances and underlying forces as key components of both the distinctive research agenda and the day-to-day business of sociology, most representatives of the discipline conceive of their task as tracking social changes as they can be captured by means of quantitative methods, at high levels of accuracy. Within the division of labor in sociology, however, the investment of time and energy required to do justice to the minutiae of concrete social change tends to be so intensive and absorbing that there is little room for constructive exchanges with those sociologists whose efforts are directed at determining and revealing the underlying forces that generated and, for practical purposes, continue to determine and maintain the face of modern society. In turn, the theoretical discourse about modern society has become so multifaceted and intricate that commitment to keeping abreast of the continually expanding literature, and especially to advancing the discourse itself, without falling prey to the temptation of engaging in undue reductionism, effectively precludes opportunities to try to tackle the concreteness of social change. If we further consider the fact that professional responsibilities and demands keep expanding, it is practically impossible, for individual sociologists to engage in a level of theoretical sophistication required to confront adequately the challenge of tracking social change.

Ultimately, to be sure, keeping up with the pace, scope, and depth of social change today should be the responsibility of the discipline of sociology as a whole, with the majority of sociologists cooperating in an agreed-upon and explicit collaborative effort. However, the proliferating fragmentation of research agendas, theories, and methodological approaches makes impossible the kind of research collaboration required to confront the complexity and contradictory nature of modern society (Gouldner 1985; Zhao 1993; Phillips 2001; Hand and Judkins 1999). As a consequence, in a manner that is similar to other social sciences, sociology has become self-referential with regard to issues, methods, and objectives, not just at the level of the discipline, and not even in terms of the divide between theory and empirical research, but within its proliferating subareas as well. Rather than illuminating the warped nature of modern social life, sociology exemplifies warped modern life in its practice and orientation.

Yet without an overarching discourse in sociology, about how the constitutional logic of modern society presents impediments to reformulating periodically the discipline's purpose and responsibilities among the social sciences, it will not be possible to confront the challenge of framing research relating to the phenomenon of "globalization." In fact, in the absence of an overarching discourse between and across all the social sciences – and especially, without determined related efforts to assert the need, and to create opportunities, for such a discourse – addressing globalization in ways that demonstrates that and how each and all of the social sciences have been and will continue to be relevant for conditions of collective life on Earth, will become increasingly difficult to demonstrate.

Towards a dynamic concept of spatial sociology

It is in this specific context that systematic sociological interest in space would provide a unique venue for scrutinizing the link between surface appearances and underlying forces. Especially for critically oriented theoretical sociologists, the tension between both levels for decades has been too obvious to ignore.[17] One of the main hurdles for sociology to play the socially relevant and enlightening role that inspired the classics in one way or another is the widespread lack of commitment among sociologists to making sure that research efforts relate directly to stated objectives. It is common practice to frame research in ways meant to be practically relevant and conducive to social improvements, or to change that is grounded in shared values. In most instances, however, researchers accept from the outset that stated objectives serve as guideposts, and are not likely to be attainable or attained, especially if they pertain to structural and systemic features of modern society. Ironically, it would not be particularly difficult to demonstrate that one prominent reason for sociologists conceding from the outset the ineffectiveness of sociological research is unwillingness to confront the actual workings of modern society, especially where it conflicts with widely accepted values and ways of reading the world. In effect, most sociologists appear to prefer avoiding the cognitive dissonance that inevitably will accompany endeavors to scrutinize the façade of modern society. Yet absent pertinent representations of the functioning and constitutional logic of modern societies, it is impossible to explain both the varying degrees of practical irrelevance of social research and its conclusions. For sociologists to explain the relative lack of relevance of social research to date, as a necessary precondition for making social research more pertinent, would require looking beneath the surface of modern society. That most sociologists refrain from doing the latter, however, does diminish explanations of the relative irrelevance of social research, and how the irrelevance is a function of the design of modern society – a profound paradox indeed.

If we were to conceive of space as the intersection between underlying forces and surface appearances, we would be in an advantageous situation, for two reasons. On the one hand, we would be able to avoid the trap of space functioning as an interpretive strategy that allows us to redefine well-established frames for sociological analysis, in the interest of opening up novel perspectives, without engendering truly innovative opportunities for meaningful and relevant research. The danger that accompanies efforts to outline potentially more productive research resulting from the introduction of a dimension of social reality that had been neglected up until now, such as space, is that doing so may create the appearance of novelty without generating truly new vistas. However, if we introduce space as the intersection between the underlying forces and surface appearances of social life in the age of globalization, the contradictory nature of the latter is included by default. Since willingness to confront contradictions is a central feature of globalization as a more pronounced extension of the constitutional logic of modern society, consideration of space would require that we thematize how concrete forms of life, as they manifest themselves in the physical world, in different kinds of structured and built environments, cannot be understood on their own terms, but as the focal point of forces at different levels of complexity in today's world (Urry 2003).

When conceived and laid out along such lines, the spatial turn does indeed constitute a step toward a kind of sociology (and concurrent disciplinary practice) that is driven by the determination to confront the complexity and contradictory nature of societal reality in the early twenty-first century, beyond the prevailing confines of power, inequality, and ideology, on grounds that point toward an updated rendering of Lemert's (2007) reference to the classics, as "thinking the unthinkable." It is quite apparent that sociology will not be able to fulfill its intellectual, disciplinary, and moral responsibilities as long as its practitioners refuse to confront the obvious challenge: to recognize that sociologists must endeavor to confront the totality of societal life theoretically, methodologically, and substantively. Economic sociology, comparative-historical sociology, political sociology, urban sociology, spatial sociology and so forth do not constitute clearly delineated areas of research, but instead are integral to the project of sociological research in a manner that must be confronted directly, rather than on the discipline's margins, by more or less recognized critics of mainstream approaches.

The need to establish a new language and new criteria for identifying and communicating the most urgent issues of our time is becoming increasingly apparent. Presently, the most common and most effective mode for framing these issues is in terms of globalism, globalization, and global studies. However, framing issues in those terms is problematic inasmuch as the emphasis is being placed on openness, opportunities, and our ability to construct the future in terms of our goals, values, and hopes. Yet it is becoming undeniable that the future will be characterized less by openness and opportunities, but by limitations resulting from the fact that the Earth is a closed and finite system, e.g., in light of the continuous expansion of the world's population, the depletion of natural resources, accumulating waste, and impending energy problems, to name just the most obvious ones.

Given its tools and orientation, especially when framed in terms of how the classics conceived of sociology as inherently critical regarding the actuality of social reality, the discipline is uniquely positioned to address related issues in a comprehensive manner, due to its foundational emphasis on the social, political, and cultural costs resulting from the pursuit of economic prosperity. Theoretically as well as methodologically, sociologists have worked on building analytical frameworks designed to frame the study of the links between institutions, practices, processes, and trends. At the same time, however, the discipline of sociology has not succeeded at leaving an imprint on society that would enable institutions and collective actors to identify the implications resulting from the imperative that societies maintain and reconstitute order and stability inherent to social structure in its specificity – including especially its spatial specificity, and how every member of society is a carrier and an embodiment of the latter – and to strive to recognize those patterns. As long as societies resist structural transformations that are a necessary precondition for actualizing to a greater extent, and over time, the power and possibility of prevailing norms and values, sociology is not in the position to enable individuals, groups, and nation-states to engage in progressively higher levels of agency.

As alienation, anomie, and the Protestant ethic (as the concrete substance of the "casing as hard as steel") are social conditions that are at the very core of modern society, individuals cannot actively overcome the limitations their prevalence imposes on efforts to construct meaningful life-histories and to make related choices. To the extent that concrete spaces channel individual and social agency, without individuals and many social scientists recognizing the link between both, efforts at agency often take the form of spatially sedimented, specific circumstances transposed directly into individuals' and society's lives. Since alienation, anomie, and the Protestant ethic first and foremost are manifest in concrete practices, relationships, and spaces, recognizing their prevalence will be necessary first steps for understanding the qualitative social changes occurring in the context of globalization (see Liggett and Perry 1995). As sociologists, one of our most important responsibilities may be our ability to conceive of, and to scrutinize rigorously, how individual identities are reflections and representations of the socially constituted features of modern society. As long as we remain oblivious to this fact, we re-enact normatively grounded practices we perceive as our very own, although they have been imprinted onto our selves in the socially embedded process of identity-formation, well before we become conscious of our own self. The link between self and society is becoming problematic to the same degree to which modern society itself is becoming problematic. As we perform our roles in concrete environments, it is virtually impossible to stay focused on the larger global context and its contradictory complexity whose paradoxical nature is more and more aggravated.

Notes

1 For an early and notable exception, see the chapter on "Sociocultural Space," in Sorokin (1964: 97–157).
2 For overviews, see Urry (2001). Important examples for sociologists working in categories of space include especially Harvey (1973), Gottdiener (1985), Lefebvre (1991), and in combination with time, Giddens (1984) – to refer just to his most important work.
3 In Dahms (2008), I have proposed that mainstream approaches start out from the assumption that it is possible to capture the contradictory, complex, and contingent nature of modern social life, without relying on 1) the concept of totality, 2) interdisciplinary research, 3) scrutiny regarding the link between society (as a concrete reality) and social science practice (theoretically, methodologically, and substantively), and 4) dialectical perspectives and tools as uniquely well-suited and necessary (though not necessarily sufficient) means tackle the dynamic nature of modern social life.
4 My reference to the systematic study of society, rather than sociology, allows the inclusion of Marx in the following discussion of the classical social theorists. During much of the twentieth century, Marx was neither regarded as a key source for sociologists, nor did his theoretical concerns influence efforts to conceive sociology as a social science until the late 1950s. (In part, this was due to the fact that Marx never saw himself or his work as related constructively to the project of sociology.) Even during the 1960s, sociologists who insisted that Marx's writings should be included in the canon of sociology, confined their reading of his contributions as a source for "conflict theory" (see Dahrendorf 1959; Coser 1956), rather than as a rigorous critic of the force of alienation, commodity fetishism, and political economy in and to modern society, as in the critical theory of the Frankfurt School. Even C. Wright Mills regarded himself as a sociologist whose concerns were more influenced and inspired by Weber's writings than by those of Marx. It was not until the 1970s that sociologists began to regard Marx's theory as part of the discipline's foundation and canon, as has been reflected in countless sociology and social theory textbooks published since then.

5 For present purpose, I will limit my discussion of classical social theory to Marx, Durkheim, and Weber. As will become apparent, there may be pressing reasons for their centrality that, as of yet, have remained undeveloped, and which may account even more strongly for the willingness to rely on their contributions, as the basis of contemporary sociology. For a "geohistorical interpretation" of modernity that informs my own perspective, see Taylor (1999).

6 It has been a bit of a cottage industry to point out all the different dimensions of social life the classics failed to recognize and scrutinize, such as nature, gender, and race. However, the more important question, especially in the present context, appears to be whether the classical contributions were *compatible with and conducive to* the study of those and other dimensions of modern social life. We must keep in mind that their efforts were directed a providing foundations for more effective analyses of a rapidly changing social environment, in ways that would translate into systematic frameworks that, in the case of Durkheim, Weber, and Simmel, were intended to engender sociology as a rigorous social science. In this sense, they were merely "beginnings" rather than authoritative canons, to be pursued, refined, reconstructed, and adapted in reference to changing socio-historical contexts.

7 My point is not that the classics successfully designed sociology, but that if their designs were misappropriated, the history of sociology may have been burdened by distorted perspectives on its purpose and responsibility. For reassessments of classics' theoretical agendas, see especially the essay included in Camic (1995).

8 See especially Steinmetz's (2005) assessments of the implicit positivism of post-war sociology in the United States.

9 Due at least in part to the neglect of Marx's contribution until the 1960s, this basic and constitutive theme of sociology as a social science has not been appreciated as fully as conditions of globalization make possible and necessary.

10 The more precise translation of Weber's related phrase in the original of *The Protestant Ethic and the Spirit of Capitalism* should have been "casing as hard as steel" (Weber [1905] 2002: 121), or a comparable rendering. One of Weber's related intentions was to communicate that the confining nature of modern existence is neither visible nor easily accessed.

11 See, for instance, Dodgshon (1998), especially ch. 7 on the built environment as a source of inertia, and therein specifically the section, "Modern societies and the social construction of inertia" (pp. 142–8).

12 For an attempt to draw out the inherently static orientation of postwar sociology, in the face of rapidly accelerating social change, see Philips and Johnson (2007).

13 For "dynamic" readings of Marx, Durkheim, and Weber, e.g. Postone (1993), Jones (2001), and Kalberg (1994).

14 It is not accidental that social scientists concerned with issues of space and time have noticed, though not developed fully, the link between Einstein's and Planck's respective framings of the analysis of physical reality, and the sociological enterprise. See Urry (2000: 106, 119–23) and Harvey (2006: 121–4).

15 For present purposes, and as an initial approximation, I am framing sociology as the social science of modern capitalist society; "modern" represents a descriptive category for societies that maintain order and stability within the force-field of "capitalism" and "democracy." Consequently, this category will continue to apply as long as the projection of the reconciliation facts and norms plays a constitutive role in the form and nature of organizations, as well as the construction of individual identities. One of the advantages that goes hand in hand with this frame is that the distinction between

modernity and postmodernity is unnecessary since, as will become apparent, *modern society* is a form of social organization that combines premodern, modern, and post-modern elements, in a manner that is inherently irreconcilable. See Schumpeter (1942), Bowles and Gintis (1986), Sayer (1991), Wood (1995), Prindle (2006), Delanty (2000).

16 See the section entitled "The Culmination of Modern Trends: Globalization as Hyper-Alienation" in Dahms (2005).

17 Which is not to say that scholars, who recognize the tensions between both levels, would also make attempts to theorize those tensions as related to the discrepancy between those levels. See Postone (2007).

7 Sex and the modern city

English studies and the spatial turn

Pamela K. Gilbert

> But all the clocks in the city,
> Began to whir and chime:
> "O let not Time deceive you,
> You cannot conquer time."
> (W.H. Auden)

Michel Foucault, in a brief essay published in 1984, announced that space, rather than time, was already emerging as a (perhaps the) primary category for critical analysis.[1] Time, the domain of narrative, had been the dominant trope and therefore category of analysis in the teleologically haunted representations of Western modernity. But space, Foucault pointed out, had also lost its sacred character with Galileo, though we perhaps had not realized it; the present era, he thought, should undertake the continuing project of its demystification. This chapter will serve three purposes: it will offer a brief survey of influential trends in the spatial turn in literature in English and cultural studies; it will sketch a historical overview of narratives of gender and sexual possibility in relation to the concept of the city, especially in modernity; and finally, it will offer some brief exemplary readings of the way those narratives have played out in literature, film, and television in late modernity.

Space in literary studies

The movement among Foucauldian scholars, in the mid-1980s, to emphasize discursive analysis and the "genealogy" of ideas corresponded with a linguistic turn in historical and geographical studies at the same time that literary and cultural studies were "thinking space" as a category of analysis. Thus the disciplines were brought closer in their concerns, though each retained its own focus and characteristic emphases. Geography was being reimagined as not simply about fixed, abstractable spatial rules, but as a discipline that would more decisively take in the social and imaginary elements of space. Even much earlier, geographers had produced careful (often Marxist) readings of space as related to relations of production, but much of it had concentrated on infrastructural rather than superstructural elements.

The Marxist influence remained strong in literary studies of space in the 1980s, particularly through the foundational works of David Harvey and Henri Lefebvre. Their work built on the strength of materialist analysis while also bringing forth a more culturally focused analysis of different kinds of space and place, and their production. In other words, they departed from the old tradition of seeing space as a material and unchanging given which is then invested with meaning by human labor and imagination, to an understanding of it as dynamically produced and "always already" imbued with meaning.

Edward Soja's *Postmodern Geographies* emphasized in 1989 that the idea of history as a way to construct stable meaning was specific to modernity; he declared that in postmodernity – specifically, the late twentieth century on – as history (time and narrative) is destabilized, it is space that comes to be the arena in which meaning is created. As narrative, with its emphasis on an understandable causality, on the coherent and universal (usually white, Western male) subject, and on a historical teleology, came to seem increasingly suspect, many theorists turned away from time and toward space – spatial relations would reveal to us a complexity and materiality which was being hidden away by narrative.

And as both Foucauldians and Marxists agree, human experience of space is always mediated by human relations with the world, material and discursive. Space is, then, not a Euclidean given; it is a materiality which we always experience both temporally and through a number of beliefs and practices. Most theorists posit two types of space superimposed on or coexisting with each other: physical space and social space (what Neil Smith calls relative space). Physical space encompasses both the natural or "given" and the built environment, and of course, as Smith points out, nature is itself produced, both ideologically and physically, through human interaction with the land, and through various scientific practices which seek to measure it. Place – the particularities of a named space experienced as unified, with clear boundaries, characteristics and a history – was often asserted as charged with meaning against the abstraction of modern space. Place could be claimed as home, as related to the construction of identity and values. From both the feminist and the Marxist sides, ideas of place were eventually critiqued – even attacked – as nostalgic mystifications of inequality and essentialism.

Edward Said contributed to a large scholarship on "othered" geographic spaces, generally read by the West or metropole as lacking or differing from the space of civilization and modernity. Race-based and postcolonial readings of space based in part on Said's *Orientalism* have focused on the spatial construction of otherness under modernity, both in the population and the landscape, on the strategies of mapping which have oppressed the colony, post-colony and so-called developing world. Said's work was foundational for scholars looking for ways to think geographic otherness historically, particularly because such otherness is often positioned as outside of time – an unchanging landscape caught in a moment prior to modernity. bell hooks and Julia Kristeva have written powerfully on the association of women with that time – cyclical, ahistorical – and on the association of women with home as a place of refuge, and of origin. This place is often associated with the mother (and later, for Kristeva, the wife), and is likewise seen as a refuge from history – this time positively as a place where identity is absolute and protected. This positive representation, however, is often at the expense of the real women associated with it, just as is the representation of other geographies as either barbaric or quaint.

But Doreen Massey has probably been the single most influential scholar in thought about space and gender, and of space-time in literary studies, a concept which recognizes that human understandings of space are also locked in temporal experiences – even a space that is statically defined geographically is temporally dynamic. As she notes, there is a tendency to regard time as "masculine, and space, being absence or lack, as feminine" (1994: 6). Massey notes that in insisting on "bounded" definitions of space or place, definitions defined against an Other, one might argue that, per object relations theory, the emphasis on "security of boundaries, the requirement for such a defensive and counterpositional definition of identity, is culturally masculine" (1994: 7). Massey argues against Harvey's separation of time and space in which Time connotes Becoming and Space, Being – this space is static, and becomes associated with the aesthetic mode for Harvey, which tends to the reactionary. Massey, however, critiqued this rejection as a characteristic overreaction – space and place could not be simply opposed, as there was no such thing as abstract empty space to begin with; there was always space-time, a field of contestatory interrelationships, out of which meanings were constantly constructed and renegotiated. Place(s) were part – and a potentially valuable, but by no means the only or ultimately important part – of those meanings (Massey 1994: 136–7).

A principal issue for gender studies of space has been the spaces of public and private, a discussion that intersects with the much studied restriction of the feminine to the domestic sphere – a concept which is associated with specific spaces such as the middle-class home – and also studies of the liberal subject and its relation to a public with influence on the political process, whether in critical opposition to a state power (as in Habermas's analysis of eighteenth-century England) or in a more complex collaboration with it. Although public and private are conceptual issues (the private citizen versus the public "man," for example), these concepts are associated with particular spaces and spatial practices (the coffee house and the public political discussion of private persons, the public square and the mob, the home and the domestic sphere separated from politics). Because of individual places' intimate economic, political, and imaginary relations to other places that may be widely geographically separated, it is more helpful to think of places "not so much as bounded areas as open and porous networks of social relations" (Massey 1994: 121). However, there is a tendency to think of both genders and places as sites of fixed identity, unchanging, with clear boundaries, and perhaps clear proprietary interests.

The modern city and its literatures

Urban space – produced out of historical moments of industrialization, population concentration, and the redistribution of space through an industrial capitalist logic – is considered particularly characteristic of modernity, with its rationalization, its abstraction, and its rigid demarcations of public and private, or privilege and abjection. Whether the city is read primarily as the site of discipline or resistance, of oppression or potential freedom, it is the urban space which is the site, *par excellence*, of literary and cultural analyses of space in Western modernity. The modern universal subject – that is to say, the liberal, implicitly male subject constructed through notions of liberty and possessive individualism, is inherently a "civil" subject, one defined by norms of civility and self-restraint as well as by freedoms to express a rather Romantic interiority. Such a subject requires a civil environment, one in which an implicit structural (legal, philosophical, though not economic) egalitarianism holds sway. In the 1970s, historical geographical analysis of inequality emerged as an important field within urban studies. In 1973, historians and literary scholars Herbert Dyos and Michael Wolff emphasized the primacy of the English Victorian city as a phenomenon in which modern patterns of urban life emerged, and a literature arose with them. Urbanization and industrialization, of course, also correlated to a rise in literacy and political activity; this emerging market, and efforts to exploit commercially and control it politically and culturally, led to an unprecedented rise in literacy and rise in cheap publications. These spatial constructs organized both the representation and the lived experience of gender, class, race, ethnicity, religion, and other identity categories that defined one's relation in terms of the universal subject.

Analyses of modern space tend to cluster around the planned city as the emblem of modernity; postmodern readings tend to emphasize the chaotic and anomic experience of that space (Frederic Jameson), or the possibilities of resistance to the centralizing, "striating" forces of modernity (Deleuze and Guattari 1987). There is a fair amount of debate over whether this period can best be identified as postmodernity or simply as late modernity. Certainly, globally, there has been a shift to late-capitalist as opposed to industrial capitalist organizations of the metropolitan cities, as the industrial city comes to be increasingly a developing-world phenomenon. Postcolonial analyses of space have focused on the image of the metropole as opposed to the colony or postcolonial space, and analyses of globalization have begun to focus on global flows of capital, labor, and information, as technologies reorganize space and a literature of exiles and migrants emerges, with characteristic tropes of nostalgia, alienation, and cultural heterogeneity expressed in spatial terms.

Having collapsed the boundaries now between what many think of as "real" and "imaginary" (or physical and semiotic) space, I should note that work in literary and cultural studies has tended in two directions: one is concerned with "actual" spaces – the space of proper nouns, so to speak (the London of Defoe, the Paris of Zola, or even Dickens's fictional but highly specific Bleak House) – and one is more concerned with a "type" of space: the city, the factory, the home. Literary studies are interested not only in how literature reflects such understandings of space – how they operate thematically and at the level of plot and setting – but also in how literature shapes the understanding of space, how it intervenes in culture to produce new understandings.

Franco Moretti has analyzed both novels and the conditions of their production to demonstrate how the dominant European novelistic fictions of the nineteenth century really emerged from three or four literary capitals and were circulated in such a way that Western literature was shaped by production from within a very few cultures. In turn, novels represented the geography of Western society in consistent ways – the space of the eighteenth- and early nineteenth-century Gothic, for example, was often in the south or east of Europe; in non-modern, non-urban sites, the location of English realism tended to be the agricultural countryside and village. As Moretti notes, a "literary geography . . . may indicate the study of *space in literature*; or else, of *literature in space*" (1998: 3); he deals with both, and then their uneven interaction. Such a study allows one to find, for example, that of sixty Gothic texts coming from England, only one is set in England; the rest concentrate their settings in Italy and France until 1800, then Germany, then, in 1820, Scotland (1998: 16). Jane Austen's stories, however, concentrate in the southern half of England – no manufacturing North, no Celtic fringe (1998: 15). He reads the spatial distribution of such tales as consolidating and finding form for the new mode of national identity formation. With Scott's historical novel, Moretti argues, geography becomes the very basis of narrative form in the nineteenth century. He then turns to two great nineteenth-century Western European literary cities: London and Paris, both of which presented crises of legibility. Some genres, such as the English silver fork novel, divided the city (us, the West End, vs. them, the east and south and pretty much everything else). Others present the city as puzzle, such as Dickens, in *Our Mutual Friend* and Eugene Sue's feuilleton, *The Mysteries of Paris*, or fields of power, as in Flaubert's *Sentimental Education*. Paris has a unique location in the Latin Quarter which allows the city to be presented as an object of desire; for Moretti, that is unique about Paris in the nineteenth century. (Of course, such a statement is a provocation – and has been treated as such!) But Moretti's point, at least in part, that the mid-century English novel tends to assert its roots in realism, but then fictionalize its subject matter in such a way that the realities of class are transcended (1998: 131), is hard to refute.

Moretti's eccentrically brilliant (and I use the spatial metaphor here deliberately) reading of literary production was followed by a more narratological manifesto, in which he proposes a history of genre through "maps, graphs and trees" in order to gain a perspective hitherto lacking. As he points out in this (non-geographical) study, however, geographical analysis of novels not only offers us the "extensive" understanding of diagramming spaces in relation to each other (Moretti gives us the example of several French novels in which the male suitors all live across the Seine from their female lovers, thus there is a spatial relationship between the characters that is consistent), but also the "intensive" understanding of the particular space – not just any space across any river, but the Quartier Latin (Moretti 2005: 55–6). Moretti finally is more interested in the extensive relationship, as he is interested in a quantitative analysis of multiple novels. But as he says "Geometry signifies *more* than geography: but it seldom signifies *by itself*" (2005: 56; his emphases). A pattern is repeated, but the pattern develops from, reflects, and sometimes creates geographical meaning.

Historically in the West, the city has been the locus of culture and civilization, though also the locus of its corruptions (in this context, set against an agricultural rural idyll). It is not until late modernity – the Romantic period and later – that the wilderness has been widely understood as a positive or idealizable space in Western culture. The city here is less a particular place than the ancient figure of the city as a zone of civility and law, within which society was structured and from which it is ruled: the classical or medieval ideal. In late modernity, the rise of the industrial city and the dominance of cultures associated with it, has created both a distinctive literature (and readership) and also a distinctive literary understanding of urban space. It participates in older visions of the city, but also creates a new opposition to an idealized rural or small settlement (village) space, which often represents purity, authenticity, and morality. Thus, the modern city is particularly complicated, representing civility and freedom, but also the fall of human beings into relations more characteristic of commodity culture than an idealized pastoral simplicity. In 1974 (translated into English in 1984), Michel de Certeau showed how the practice of space, the experience of following itineraries within the urban space, was a process of reading, perhaps even of writing, and the city's urbanity is often portrayed as a complex text of which one either masters or finds overwhelming and illegible. The literature of the modern city and its public spaces "is replete with descriptions of boulevards and cafes, of fleeting, passing glances and of the cherished anonymity of the crowd" (Massey 1994: 233).

Urbane pleasures

This concept of "urbanity" still informs representations of the city, in literature, sociology, and planning. However, its shadow – the threat of the specifically urban predator, enabled by the density and anonymity that makes civic freedom uniquely possible – has been an important theme in representations of the city from about the late eighteenth century onward, and has occasioned moral panics sporadically since then. For women, dangers are encoded as victimization or prostitution, whereas the pleasures of the city, sexual or otherwise, are proffered through consumerist imagery. Masculine dangers are bodied forth either in the male as predator – sex criminal, hunter – or in the male as victim of seduction by prostitutes as consumer objects. When male sexual agency is celebrated, however, the city is offered as a site of freedom through anonymity and consumption, again, often of prostitutes as consumer objects.

Theories of the city which celebrate this mastery, and especially of the literary and artistic city, are often theories of *flaneurie*, derived from Benjaminian models, and developed through the image of the *flaneur* as of course male, then explicitly middle- or upper-class and urban, focusing on the gaze as the primary mode of intersection between *flaneur* and both others and consumer goods. The *flaneur* does not necessarily buy; he consumes through his eyes, and enjoys his capacity to be in, but not of, the crowd – or of the crowd only by his own volition. The intimacy of the crowd is without obligation; intimacy ultimately reinforces the *flaneur's* isolation. Classically, for example in Baudelaire, Benjamin's chief example, the *flaneur* is enchanted by glimpses of anonymous women (see, for example, the poem 'À une passante' or "To a Woman Passer-by"). Deborah Epstein Nord has done much to elucidate the role of the *flaneur* in London. As she observes, the *flaneur* was assumed to be masculine (and heterosexual); the lounging, loitering woman either was, or risked being identified as, a prostitute (Nord 1995: 11).

The ultimate *flaneurs* in London literature of the earlier nineteenth century were Tom and Jerry, in Pierce Egan's 1821 *Life in London, or The Day and Night Scenes of Jerry Hawthorne, Esq. and his elegant friend Corinthian Tom in their Rambles and Sprees through the Metropolis* – a combination of journalistic and fantastic travelogue and picaresque focusing on extremes of urban London. For example, they pass from Almack's, the famous upper class ballroom and marriage market, to "All Max," a seedy bar in the East End populated by a racially mixed, boisterous crowd of dancers, prostitutes, and drinkers (Nord 1995: 30–2). Still, Nord recovers a few women who narrate the city. Her early nineteenth-century example, Flora Tristan, who wrote *Promenades through the Streets of London*, views prostitutes, as do her male predecessors. But as Nord points out, for Tristan, this cannot be expressed simply in terms of lust or pity, but must bear the burden of identification and fear of assimilation; as a woman, Tristan recognizes her likeness to the sexually exploited woman, and her vulnerability. It is only her male companions who protect her from insult or violence. Therefore the *flaneuse* does not experience urban space as freedom in the same way as it is represented for male characters or authors. It is not until the 1880s that Nord identifies women for whom the streets of London represent a site of agency and freedom; and for these New Women, it was often through a sense of mission, as social workers, that they celebrated this sense, rather than simply as disengaged consumers of spectacle. As Massey observes,

> The spatial and social reorganization, and flourishing, of urban life was an essential condition for the birth of a new era. But that city was also gendered. Moreover, it was gendered in ways which relate directly to spatial organization. . . . This period of the mid-nineteenth century was a crucial one in the development of 'the separation of spheres' and the confinement of women, ideologically if not for all women in practice, to the 'private' sphere of the suburbs and the home . . . The public city which is celebrated in the enthusiastic descriptions of the dawn of modernism was a city for men . . . The women who did go there were for male consumption.
>
> (Massey 1994: 233–4)

Though the *flaneur* is, as I mentioned earlier, generally heterosexual, the lounging male gazing at others has the capacity to inspire fear, not only among vulnerable women, but of his capacity for homosexual contact. The association of homosexuality with both the promiscuous contact of the city and its anonymity – its privacy paradoxically located in its public nature – allow both for the visibility of alternative practices and fears of their invisibility "in plain sight" – that the coded behaviors and gestures of such men in public may mean that queer sex may be taking place in the next stall in the public men's room. Walt Whitman's (1819–92) "City of Orgies," from *Leaves of Grass*, is specific about the pleasures of Manhattan: "City of orgies, walks and joys," he apostrophizes,

> City whom that I have lived and sung in your midst will one day make / Not the pageants of you, not your shifting tableaus, your/spectacles, repay me, / . . . / Not those, but as I pass O Manhattan, your frequent and swift flash of eyes offering me love, / . . . / Lovers, continual lovers, only repay me.

The "city of walks" offers itself to the *flaneur*, but ultimately, it is not the pleasure of spectacle that he seeks, but that of (however multiple or fleeting) intimacy. In 1929, Garcia Lorca would, walking by the East River, where he sees boys working and singing, and thinking of the poet who identified the city with sexual possibility for him, mirror Whitman's negatives and double them, superimposing Whitman's "virile beauty" over the "mire and death" of the industrial Bronx: "Not for one moment, beautiful old Walt Whitman, have I ceased to see your beard full of butterflies, nor your shoulders of corduroy worn down by the moon, nor your thighs of virginal Apollo, nor your voice like a column of ashes: ancient and lovely like mist," even though none of the boys stopped: "None of them wanted to be a cloud. None of them sought for ferns."[2]

Although queer sex has always been associated with the city in Western literature, there are some cities more than others that are associated with such pleasures. The metropolis, always, as a large mixing ground for all tastes, is thought to provide such opportunities. But the colonial city, especially the Asian or north African city, is particularly associated in European literature with queer sex (Robert Aldrich mentions Saigon and Tangiers, at different historical moments), as are the historically charged premodern European cities of Venice and Athens. As Aldrich mentions, it is often authors (and filmmakers) who cast the city in its role; Paul Bowles and E.M. Forster have done much to give a European and North American population its ideas about the East (2007: 90). Not coincidentally, such cities are also associated with the sexual marketing of subaltern populations. Whereas London's Soho and San Francisco's Castro are distinguishing features of those cities, they do not define the Western view of those cities in the way that Asian and African cities are defined for English-speakers through notions of erotic exoticism and sexual tourism. Nor is this phenomenon limited to Anglophone cities; Paris "has" the Marais, not the other way around.

Dangerous liaisons

Foucauldian scholars Thomas Osborne and Nikolas Rose describe the late modern liberal ideal of urban space as that of "virtuous immanence." The city is an arena for citizens to exercise their freedom to self-govern, for competing desires to assist in the emergence of the perfect society from subjects' experience of the city itself. But the freedom of urban life also led to another form of immanence: the "vice, rebellion, insubordination" of the mob and its constituents (1998: 3). This created conflict thematized spatially in the open visible space of the modern city against the resistant, dark recesses of the premodern city the modern planner tried to rationally restructure. Peter Stallybrass and Allon White have argued that the nineteenth-century city in England is organized around the binaries of filth/cleanliness or impurity/purity and the fear of their transgression (1986: 136). This fear was articulated through the "body" of the city, which had to be scrutinized and controlled (Stallybrass and White 1986: 125–6). As I have argued elsewhere, this surveillance – equated with the very essence of civilization – was institutionalized through sanitary inspection and had entered both literary and visual culture, the latter principally in the form of maps (Gilbert 2004). By mid-century, the "lower bodily strata" of both city and its inhabitants that Stallybrass and White describe being identified with both sewage and underclass behaviors was increasingly thematized as disease, vice, and anti-modernity. This trend is strong in both literary and non-literary discourses such as urban planning, legislative politics, evangelicalism, and medical writings. Doctors spoke of the incestuous practices of those who lived too close together in the city's slums; clergy spoke of the dark courts and tenements in purgatorial or infernal terms as sites of godlessness expressed in precocious and illicit sexual practices: prostitution, sexually active children, sodomy and promiscuity. Sexual transgression is the disorder that launches narratives, makes game of the public–private divide, and disrupts civic order. Sex, except when decorously reproductive and restrained within doors and within marriage, is positively uncivil.

The association of the city with sexual license and perversion is an old one in Western culture, going back to the Old Testament and the literature of ancient Rome. In the USA, the image of the utopian, gleaming, and specifically New World 'city upon a hill' from Matthew 5:14 was made famous in Puritan John Winthrop's famous 1630 New England sermon, "A Model of Christian Charity." It has been repeatedly invoked since then, mostly notably in recent decades in Ronald Reagan's speeches, to describe the perfectly civil, utopian site of freedom and purity – indeed of citizens so pure that their very freedom only enhances their purity rather than allowing them occasions to stray from it. In 1974, Reagan quoted Winthrop and insisted that America had fulfilled the promise of the city of the hill: "We are not a sick society. . . . We are indeed, and we are today, the last best hope of man on earth." This image of godly and healthy American civility is often set against the image of the "cities of the plains" defined specifically by sexual deviance and "cleansed" by God in a cataclysmic act of destruction.

It is non-normative, non-reproductive, sexuality which is most closely associated with disgust and boundary transgression. Sex shops and their products, for example, are regulated through zoning and obscenity laws, although many people do not find the possession of such products offensive. But for visceral horror that unites a large majority of the public, nothing competes with sex offenders. Over the last two decades of the twentieth century, a barrage of notification laws (often collectively referred to as Megan's laws) have aimed at making visible and limiting the mobility of sex offenders within high density areas, "mapping" the contaminating threat posed by such individuals. By 1999, all 50 states of the USA had passed some such legislation (Filler 2004).[3] Such laws publicize the home addresses of sex offenders, and often require that such persons live in a certain spatial relationship to "vulnerable" sites – keeping them 1,000 or so feet from schools, churches, or day care centers. Several cities, because of the density of population and therefore proscribed sites, have been forced to move their sexual offenders in high concentrations to nearby rural areas or leave them homeless, prompting offenders to drop out of police surveillance programs.[4] These laws offer to neutralize the dangers of the urban landscape by making them cartographically visible, but this project reveals more about our representational relationship to sex and urban space than it does about the management of sex crimes.

Steve Macek notes that "panic over inner-city pathology and chaos has been structured . . . by a conservative discourse that . . . constructed the central city as an object of middle-class fear" (Macek 2006: p. xvii). As scholars such as Steve Macek, and Stallybrass and White have pointed out, since at least the nineteenth century, the poor of the urban core have been associated with and ultimately identified as moral and physical impurity. Physical filth has been culturally defined in the West as related to the abject materials of a transgressive physicality – "matter out of place" in Mary Douglas's phrase – including feces, semen and other bodily fluids. Moral filth has often related to proscribed sexual practices, and various urban horror stories attest to this, such as Jack the Ripper and murdered prostitutes in a Jewish East End slum of London. As Judith Walkowitz argues, "the prostitute was the quintessential female figure of the urban scene, a prime example of the paradox, cited by Stallybrass and White, that 'what is socially peripheral is so frequently symbolically central'" (1992: 21). "A logo of the divided city itself," the prostitutes were opposed to the "classical elite bodies of female civil statuary that graced the city squares" (Walkowitz 1992: 21, 22), and the Ripper was the extreme example of the more mild sexual dangers of being "spoken to" that middle-class girls risked going into Oxford Street or Picadilly unaccompanied (Walkowitz 1992: 52).

Seth Koven follows Walkowitz's work to show that the outreach or social work performed by middle- and upper-class women and men in working- and lower-class areas was understood through narratives not only of sexual danger but of sexual opportunity and homosexuality, as well as deviant homosociality that might challenge the dominant orders. The East End, like St Giles in the West End at mid-century, was associated not only with racial otherness (the Jews of the East End, the Irish of Seven Dials, the Asians near the Docks), but with sexual otherness. These categories, of course, overlapped. As Walkowitz elaborates, feminist responses to the Ripper capitalized on women's vulnerability in the city, and thus unconsciously collaborated with repressive structures reinforcing women's exclusion from the equal enjoyment of public space and the pleasures of the city. It probably also contributed to the continued understanding of those pleasures as explicitly sexual today.

Urban space has been defined in late modernity largely in terms of spatial proximity of the polluting and the pure. Joyce Carol Oates has posed the problem this way:

> the City, an archetype of the human imagination that may well have existed for thousands of years . . . has absorbed into itself presumably opposed

> images of the "sacred" and the "secular." . . . A result of this fusion of polar symbols is that the contemporary City, as an expression of human ingenuity and, indeed, a material expression of civilization itself, must always be read as if it were Utopian (that is, "sacred") – and consequently a tragic disappointment, a species of hell.
>
> (1981: 11)

What is most pure is also most subject to defilement. The figure of the child-molester guarantees the coherence of sexuality and the law by being the impure object of fear, desire, and discipline. The most successful of the *Law and Order* (1990–present) television show franchise, the *Special Victims Unit* or *SVU* (1999–present), focuses on the commission of sex crimes. The public is fascinated by pedophiles, and they are particularly associated with the urban setting.[5] As opposed to the gothic machinations of the *X-Files* (1993–2002), which often played out in rural and remote locations, Americans' favorite crime stories are set in meticulously realized urban settings. *Law and Order* carefully sites its crimes in specific New York City settings, and the *Crime Scene Investigation*, or *CSI* (1999–present) franchise is associated with particular cities, e.g., *CSI Miami* (2001–present). The obsession with mapping and containment shown in zoning and offender notification laws provides a way to make fears associated with modernity legible through narratives (and cartographies) of urban space. The *SVU* series is shot in New York on location, and emphasizes local color. New York City mayors Rudy Guiliani and Michael Bloomberg have both appeared on the show, blurring the lines between fiction and reality, and underscoring the city's dangerousness, as well as the mayors' commitment to civic order. In 2004, a road leading to the Chelsea Piers site where the series is mostly filmed was renamed "Law & Order Way." Clearly the pleasures of the series are linked to the pleasures and dangers of the city.

This formulation, of course, is not specific to America in the late twentieth century – Fritz Lang's 1931 film *M* (a more complex portrayal of the child-killer as the ultimate danger of urban space) associates him with the underside of Berlin, and initially it seems that all the forces that pursue him, even the other criminals (who are after him because the police attention to the city is making it too hot to hold them), become a policing force. As Edward Dimendberg points out, when the criminals meet to discuss their plan, the map enters the *mise en scène* as a powerful symbol of visibility and rationality. But the criminals and terrified citizens alike are also the force of the mob, and the fable of a rational city of light against a force of darkness is undercut by the complexity of Lang's portrayal. Overhead shots of the city streets at night offer a fantasy of visibility, but the most frightening scenes play out in generic dark basements and alleyways, where visibility is limited. Lang reflects back to the viewer his or her own frustrated desire to see and control; at the end the trapped criminal tells the mob/jury they cannot understand what it is like to be him, and Lang dramatizes the impossibility of this knowledge, of the fantasy of the knowable city and the knowable self. Dimendberg notes that the murderer, Beckert, is fascinated with consumer goods – he is often seen staring into store windows. He is the very type of the *flaneur* who consumes females as he consumes toys; the sex-criminal is the *flaneur* turned vicious, who consumes their actual lives rather than their appearance as an aesthetic experience. However, the figure of the prostitute always mediates these two possibilities: the woman on display, for consumption, and the woman consumed, destroyed. The woman who is sold is always already "ruined," a vampiric figure or living death allowed or doomed to wander for a while, but already marked for destruction.

The city is particularly associated with women's seduction into prostitution. Nineteenth-century British artist and poet Dante Gabriel Rossetti's painting *Found* (which he began in 1853, and continued to work on for almost three decades) shows the country girl in her prostitute's finery, groveling before her abandoned beau in his laborer's smock, as he discovers her in a visit to the city of London. Rossetti describes it:

> The picture represents a London street at dawn, with the lamps still lighted along a bridge that forms the distant background. A drover has left his cart standing in the middle of the road (in which, i.e. the cart, stands baa-ing a calf tied on its way to market), and has run a little way after a girl who has passed him, wandering in the streets. He had just come up with her and she, recognizing him, has sunk under her shame upon her knees, against the wall of a raised churchyard in the foreground, while he stands holding her hands as he seized them, half in bewilderment and half guarding her from doing herself a hurt.
>
> (Cited in Nochlin 1978: 139)

The girl is assimilated to the street, the gutter, and death (the churchyard which Rossetti's sketches indicate he intended for the area by the wall); the laborer stands framed by the sky, bringing his own cattle to market (bound and helpless in the cart). The lamb, symbol of gentleness, helpless immaturity, and Christian sacrifice, is to be cared for by the shepherd, but lambs are also sold; the city is where the gentle shepherd enters the logic of the marketplace, and where domestic innocence is corrupted out of doors'. The bridge is over the Thames, and the stereotypical end of fallen women was suicide by drowning – as can be seen in the third panel of the famous 1858 Augustus Egg triptych, *Past and Present*, in which a fallen woman ends her story homeless under a bridge in London (under a "Found Drowned" poster), contemplating the water as she holds her illegitimate child.

"Jenny," Rossetti's poem on the topic of prostitution, also takes as its topic the familiar story of the innocent country girl who comes to the city, not knowing what it may be, and falls:

> Jenny, you know the city now.
> A child can tell the tale there, how
> Some things which are not yet enroll'd
> In market-lists are bought and sold
> . . . And market-night in the Haymarket.
> Our learned London children know,
> Poor Jenny, all your mirth and woe;
> Have seen your lifted silken skirt
> Advertize dainties through the dirt;

He associates her ruin with a specific geographic site (the Haymarket, a portion of the theater district associated with prostitution in this period), and with the "market" and the mid-century development of advertizing. He muses that there is no point in speaking to her on this matter, for her mind is already destroyed, identical with the sewage-defiled drains of the city, and therefore too polluted to hold an idea clearly: "For is there hue or shape defin'd/In Jenny's desecrated mind,/ Where all contagious currents meet,/A lethe of the middle street?" The body of the ruined woman becomes the polluted streets of the city – her presence in this public space, outside of the domestic sphere, means that she is assimilated to the dark side of the urban itself. The confusion of night and day, the darkness, and the inability to discern "hue or shape" identifies Jenny's mind with the abject city: obscure, dangerous, unknowable.

Rossetti selects the Haymarket because it was associated with prostitution. But the history of the neighborhood, which contributes to its present status, is also germane; after all, Rossetti could have placed Jenny in any of the many (and there were indeed many) locations known for prostitution. The Haymarket – historically called so because it was the site where farmers brought hay to market thrice weekly – also formed a liminal space between the privileged site of the late Renaissance court (St James) and the more rural St Martin's in the Fields. The location of the court transformed this area, making it a meeting point for people of different classes, a theatrical area with an opera house (but also an area known for freak shows in the early eighteenth century), and therefore a gathering place for performers, their patrons, and their hangers-on. It was also a gathering place for foreigners. Wealth and poverty, local and international, commingled in the Haymarket. In 1830, the market became a traffic problem, as the Haymarket area was increasingly urbanized, and so the farmers' market was moved. The area was transformed entirely into a residential and entertainment district. But the theme of the farmer coming to market his wares and the farm girl coming to be marketed herself ties the urban location of the Haymarket in "Jenny" to the apparently more peripheral location in *Found*. The resonance of the (for Rossetti) exploitative relation between urban and rural, between the wealthy west side and the rural maiden with no capital but her own body echoes in his emphasis on the theater district as an area wherein one still brings one's wares to market on a Saturday night. The Haymarket was between St James Parish and St Giles – often invoked as shorthand for the extremes of wealth and poverty in close proximity in London. The two parishes both began as rural leper colonies, and the founding of St James' Palace lifted one to wealth and status as the city grew to encompass both parishes. St Giles, on the far side of the theater district from St James, became a byword for criminality and disease, the worst slum in the West End. The archetypical trajectory of a fallen woman might well be traced from the country to London – and within London, as she aged or became ill, from St James to St Giles.

But what of the woman who successfully sells her sexuality? Dreiser's Carrie Meeber (*Sister Carrie*, 1900) is seduced by a salesman she meets on the train on the way to Chicago from small-town Wisconsin; next, her liaison with the banker Hurst takes her to New York and Montreal, where he descends to the depths of the urban experience and she ascends to the heights of the stage. The consumer desire that she manifests in her encounter with the new experience of the urban department store early in the novel is represented by Dreiser as in some sense a positive force – it gives shape to a desire for a larger, more perfect existence which is linked to Carrie's artistry. It is beauty which the city offers her, and finally, that offering is her own beauty.

> Carrie passed along the busy aisles, much affected by the remarkable displays of trinkets, dress goods, stationery, and jewelry. . . . The dainty slippers and stockings, the delicately frilled skirts and petticoats, the laces, ribbons, hair-combs, purses, all touched her with individual desire. . . . She realised in a dim way how much the city held – wealth, fashion, ease – every adornment for women, and she longed for dress and beauty with a whole heart.

This kind of "successful" exploitation of sexual possibility, however, is figured entirely in terms of consumer desire. There is never any evidence that Dreiser's Carrie has sexual desire for any of the men who court her. She desires consumer goods, and she desires to become one herself – an object of display on the stage. Carrie enters the world of luxury as much as a commodity as a consumer; she consumes in part so she can become a more perfect commodity. Consumer, victim, or agent, such women figure largely as *femmes fatales* in narratives of the city; either they can be rescued and redeemed, or they draw their hapless male victims – often unintentionally – to their own demise, as Hurst ends his days in the landscape of urban Gothic, an alcoholic wreck.

The pleasures of the city for women are often associated with consumer desire, and sexuality plays out through those metaphors. The dangers are associated with sexual corruption, exploitation, and violence. When women enter the market, bad things happen to them; when men enter the market, it is a site of opportunity and pleasure. For men, the city is presented more often as a site of freedom to consume women, but also as a site of release from traditional sexual roles – perversion as opportunity. Thus the city's dangers are part of its pleasures; the abject sites of *noir* are presented as touristic sites to be consumed themselves. The dangers are figured as moral and health related – less as exploitation than corruption. In the works which reverse this dynamic, and position women as empowered to enjoy the pleasures of the city, sexual and otherwise, the sites of abjection tend to be purged; the city becomes an amusement park in which slums, sites of racial otherness, or areas of industrial ugliness or urban waste space do not figure.

In the late twentieth century, even though the urban space continues to be fraught with sexual dangers, especially for women, it is also the site of new freedoms. New genres have arisen to chart these so-called "postfeminist" stories. The television show *Sex and the City* (1998–2004) modifies its celebration of sexual freedom for its four characters by ultimately encompassing their sexual quests within traditional narratives of heterosexual monogamy and romantic love. For all that, its many years of exploration of sexual themes including fetishism, homosexuality, promiscuity, and celebration of possible female pleasure in those "transgressive" acts maps a city of pleasurable possibility and sexual agency as specific as the city of sexual danger mapped by *SVU*, even if it presents those pleasures in the same consumerist terms as those nineteenth-century narratives alternatively excoriated and chronicled them. The official website for *Sex and the City* offers an interactive map of Manhattan – "50 memorable Manhattan locations" – based on particular episodes and organized thematically; there is a geography of shopping, romance, hotspots, landmarks, and "awkward" (designating embarrassing events in the characters' adventures). It is to be noted that the map cuts off on the north side in mid-Central Park. Harlem does not exist. This pleasurable city is, with few exceptions, the site of whiteness and affluence; even Miranda's exile to Brooklyn upon her marriage (close to the end of the series), seen as a kind of suburban site of loss of freedom and accession to adult responsibility, parenthood and monogamy, does not evoke class or ethnic diversity. This "City" is purged of sexual danger through the visual erasure of the urban landscape associated with *noir*. We never see basements, alleys, or urban filth in *Sex and the City*, and night scenes are always vibrantly lit. The most transgressive character, Samantha, is a momentary exception, as she is seen in a loft in a "transitional" or gentrifying area – the meatpacking district – which puts her at odds with the black and Latina transvestite prostitutes who cluster outside her window as she tries to sleep. But the scene of their confrontation is played for laughs and resolved with a cocktail party – there is no real threat of violence. We never see this site again in the series, and it is never presented in *noir* terms – as gritty, dirty, or truly dangerous.

Still the iconic sequence in the opening credits in the 1998 season implies otherwise, as it plays knowingly with traditional urban tropes of sexual vulnerability and objectification: Carrie walks through the streets in a fluffy white tutu; we see, as from her point of view, iconic buildings (the Chrysler building, for example), city streets, and blue skies. She is enjoying the visual pleasures of the city, and simultaneously enjoying the fantasy of ballerina-like performance and perhaps, purity. Suddenly a bus bearing her image drives by in the dress she first wears to bed the series' elusive male romantic lead Big ("Carrie Bradshaw knows good sex," reads the logo), and the bus splashes her pristine skirt with dirty water from the gutter, stopping her euphoric progress in its tracks. All the elements are there – the freedom, the pleasure, the consumer items (the dress), and the danger of contamination from sexuality, figured as the dirt of the streets, just as it is in nineteenth-century representations. But the new element is that the woman seeks sex, rather than simply being sought. Sex, here, is both that which offers and defiles freedom and pleasure, with the doubling of the woman as subject (Carrie, the viewer or *flaneuse*, in a virginal tutu) and object (Carrie in a sexy dress on the side of a bus). There is perhaps also some reference here to the episode in which Carrie stands with her friends and is distressed as they all see the same image on the side of a bus, but with sexually explicit graffiti inscribed on it. Once offered for public consumption, the sexual woman no longer controls her body.

If, as Moretti suggests, every genre requires a particular space, without which its existence is impossible, its story cannot be told, then it is probably safe to say that the story of women's sexual as urban consumers cannot be told in the Gothic or gritty realist space of urban narrative as constructed in Dickens or Gissing. Unlike, say, Arthur Symons's typical narrator, who is a connoisseur of street-walkers and aestheticized the particular beauties of London poverty,[6] Carrie Meeber's success consists of being lifted out of the sites of poverty and not being exposed to them again. Dreiser's realism requires both that we see the sordid city and see that Carrie's desire is to escape it. Bridget Jones (of the film and popular novel by Helen Fielding, *Bridget Jones's Diary*), the women of *Sex and the City*, and other "chick-lit" characters cannot be imagined at any point in their trajectory within a slum setting – the heterosexual pleasures of the city are generally merely sordid or comic when taken by women in poverty and only tragic in sites figured as dangerous. Bridget could no more live out her narrative in West Ham than Samantha can be imagined in East Harlem. On the other hand, the *noir* space of the city as represented in late nineteenth-century realist narrative and in film becomes the location of sexual victimization and (sometimes) queer pleasures. These spaces, of course, coexist and overlap – the brightly lit consumer space and superstructure of shops and highrises and the industrial space of production and infrastructure (warehouses, docks and alleys) require each other's existence in order to exist themselves. But once sex enters the narrative, they cannot be acknowledged in the same plot except as the tragic upheaval of the repressed into the space of civility – *Sister Carrie* or *SVU* or *American Psycho*.

An exception is queer pleasure, sometimes related to the marginality of its spaces. As many scholars have observed, male homosexuality is often associated with the abject spaces of both the body and the built environment: public toilets, alleyways. Although sociologist Laud Humphreys's famous "tea-room trade" study (1970) positions such spaces as sites of community, recognition (less of the person than of desire), and pleasure, part of their pleasure is their danger, both of discovery and of potential exploitation (and now, many years later, of disease). But this is less frequent in representations of women as agents of sexual action than in those of men (*The L Word* is homo *Sex and the City* in LA), whereas writers like John Rechy (*City of Night*) enter a long history of representations of male homosexuality and prostitution in urban space. Though these are rarely found in mainstream or canonical novels before the late twentieth century, the tradition considerably antedates what we identify as belles-lettres. Morris Kaplan quotes The *Yokel's Preceptor*, a London guidebook from the 1850s, that warned the rural visitor of the dangers of

> these monsters in the shape of men, commonly designated margeries, poofs, etc., of late years, in the great Metropolis, renders it necessary for the safety of the public that they should be made known. . . . Will the reader credit it, but such is nevertheless the fact, that these monsters actually walk the streets the same as whores, looking out for a chance!

Such "monsters" could be found by the diligent reader in many places; the book specifies "the Strand, the Quadrant, Holborn, Charing Cross, Fleet Street, and

St. Martin's Court," and identifies the nineteenth-century equivalent of toe-tapping in an airport restroom: "They generally congregate around the picture shops, and are to be known by their effeminate air, their fashionable dress. When they see what they imagine to be a chance, they place their fingers in a peculiar manner underneath the tails of their coats, and wag them about – their method of giving the office" (120–1, cited in Kaplan 1999). An 1881 British erotic novel, *Confessions of a Mary-Ann*, places its protagonist picking up men in Piccadilly, near Cleveland Street, the site of the famous telegraph boy prostitution scandal of 1889 (Kaplan 1999: 283–9). In both examples, it is evident that the locations of the heterosexual sex trade and the homosexual one are overlapping or identical. Although this is not always or necessarily the case, there is an identification of certain zones as liminal or corrupt and associated with sexual commodification and deviance.

Another exception is also associated with the marginal zone of the immigrant community, the uneasy place of the subject who comes to the city for a new life, but finds access to metropolitan identity complicated and often simply impossible. Although many such novels chart a story of disillusionment and loss, such as Upton Sinclair's 1906 *The Jungle*, in which Ukrainian immigrants seek a new life in Chicago, or V.S. Naipaul's (1967) *The Mimic Men*, whose Sikh subject idealizes London until he lives there, sometimes the metropolitan city is constructed as a site of freedom and possibility even after the disappointment and struggle of the new immigrant: this is particularly so for women or other subjects who are subaltern in their own culture. Monica Ali's *Brick Lane* (2003) charts the life of Nasneen, who comes to the iconic Tower Hamlets location in London through an arranged marriage to a man much her senior. Although she comes to respect him, when he returns to the place he, after an adult lifetime in London, still thinks of as home, she refuses to leave London. While he idealizes East Pakistan/Bangladesh, she thinks of her sister, Hasina, who leaves the village only to be trapped and destroyed in the Bangladesh capital city of Dhaka, first by her violent marriage, and then by those who exploit her status as an unprotected woman – and then she thinks of her London-born daughters. Although she has found London difficult and hostile, she has also made significant bonds with other women and the younger generation – her children and her young lover (who is also disenchanted with London and reacts by turning to fundamentalism). Whereas Naipaul's disappointed protagonist muses,

> It is with cities as it is with sex. We seek the physical city and find only an agglomeration of private cells. In the city as nowhere else we are reminded that we are individuals, units. Yet the idea of the city remains; it is the god of the city that we pursue, in vain,

Nazneen discovers sexual fulfillment in London with a lover who claims a Bengali identity, but stammers when he speaks Bengali, though not in English. While Hasina lives the nightmare cautionary tale of the naïve woman who goes to the city and is destroyed, Nazneen is ultimately empowered by her painfully acquired mastery of the foreign language and culture of London.[7]

Ali shows the difficulty of the metropolis for the postcolonial subject, but also the ways in which some subjects, subalterns in their own community, can carve out forms of opportunity and identity unavailable to them in their place of origin. (Though it should be noted that a number of Tower Hamlets' residents found Ali's novel offensive, and threatened to burn it in public protest.) The Pakistani Muslim area of Brick Lane has been the setting of much literature and is synonymous with London Sylhet culture and food (the Brick Lane Curry Houses on 2nd Ave in New York advertises that "the next best curry house is across the Atlantic"); in the late nineteenth century, it was Whitechapel Lane and synonymous with Eastern European Jewry, and its mosque was then a synagogue. (In the early nineteenth century, the area was largely Irish; in the eighteenth, it was populated by Huguenot weavers.) Today, Brick Lane is a trendy tourist attraction, filled with Sylheti restaurants, coffee houses, nightclubs, and galleries, in the midst of a gentrifying but uneven area of London Muslims. The Lane is largely patronized by West-Enders and foreign tourists, but is claimed as a place of pride by the local community.

Despite its changing history, it retains a certain "geometric" relation to wealthy West London – it is always a place where not-particularly-welcome immigrants congregate and practice their minority religion and culture, even today when it is in the heart of the city instead of at the margin (the east "Aldgate"). Thus it is a place that is very distinctive, though it is not distinctive in exactly the same way over time. (It remains to be seen whether now that the neighborhood has been so thoroughly absorbed into the heart of the city, it can still be a site of entry for othered groups; it would seem that this point of entry is largely moving to the edges of the city – south of the river, to Brixton and Lambeth, for example.) Further, it is a place that could not exist outside of a great modern, and probably metropolitan, city – a site that is or has been the center of empire (and thus, inevitably, of postcolonial imagination, desire, and resentment) and large enough to support the complex economic and cultural activity required when a large number of people from the same (often) rural areas with largely the same marketable (or unmarketable) skills and knowledge come at once to be absorbed into an alien "local" economy. Size does matter, and so does history – with all the prejudice and sometimes violence that marks the immigrant's accession to participation in the metropolitan core, the history of similar shocks and the diversity of past participants in that process provide itineraries and strategies, both for the existing urban population and the immigrant, to "make their way." Ali boldly claims a place for her novel as "the" immigrant novel – or at least the Sylheti novel – by identifying Nazneen's story as that of Brick Lane.

City stories

City of pleasure, of danger, where all things have a price. Such portrayals are somewhat geometric in Moretti's terms – the City is iconic for every modern Western city (London, NY, Paris) and the spatial relations are largely tied to simple oppositions (uptown, downtown, West End, East End, etc.). But they are also what Moretti would call "intensive" in particular ways. Dickens uses the Thames in *Our Mutual Friend* to figure the circulation within the body of the city, and to mobilize the trope of rebirth, as he might have used the Seine in Paris. But he uses it as an emblem for pollution and defilement as only the Thames could be used – an estuary polluted by sewage that then re-entered the city twice daily with incoming tides, referring to geographic, engineering, and medical texts of the day. The Village can be a site of sexual freedoms for the four protagonists of *Sex and the City* in part because it is well known as a homosexual community – and therefore, anything the characters do is offset in its transgressiveness by its heterosexual normativity. But the women do not settle down there; they move through but do not put down roots in this liminal space.

The existence of metropolitan centers continue, as Moretti shows the European West did in the nineteenth century, to make possible the disproportionate production and thus control over representations of the urban and of geography generally; it also remakes the possibilities of populations, spreading a few languages and cultural value systems worldwide, establishing itineraries for global migration, imposing hegemonic genre and form on the local production of culture that attempts to speak to a broad audience. But it does so unevenly. The objects of this representation look back and write back, even if they perforce do so in the metropolitan tongue. Even if the tendency of metropolitan literatures in late modernity has been to reinforce high-modern understandings of space, narratives chart temporal-spatial paths that map, unmap, and remap our Brick Lanes, our Greenwich Villages, our West Ends, and Fifth Avenues. And as it ever has in modern literature, sex disrupts the orderly progress of time and violates the decorum of spatial boundaries. Walt Whitman mused,

> ONCE I pass'd through a populous city, imprinting my brain, for future use, with its shows, architecture, customs, and traditions;
> Yet now, of all that city, I remember only a woman I casually met there, who detain'd me for love of me;

This woman has been variously identified as a metaphor or a prostitute, female or male, mulatto or black or white, in New York or New Orleans. Sexual transgression is the uncivil heart of modern life; that boundary line of which the transgression by a few guarantees the possibility of the many's civility. Finally, what remains of the city, its architecture and traditions? A site for the refusal or acceptance, the attempt or restraint of a transgression: an idea, a poem, a passionate intersection of narrative and place, time and space.

Notes

1 Foucault advanced this idea in 1967, in a lecture given in Berlin, but it was not published until 1984 in France, and appeared in *Diacritics* in English translation in 1986. In this, although he is so often perceived as a point of origin, he was explicitly doing homage to Bachelard; but for literary critics in the Anglophone world, our story begins

more practically with Foucault (and spatially, it hardly matters, as it places our origin firmly – and with characteristic eccentricity – in France).

2 'Ni un solo momento, viejo hermoso Walt Whitman, he dejado de ver tu barba llena de mariposas, ni tus hombros de pana gastados por la luna, ni tus muslos de Apolo virginal, ni tu voz como una columna de ceniza; anciano hermoso como la niebla . . . Pero ninguno se detenía, ninguno quería ser nube, ninguno buscaba los helechos'.

3 For these figures and an extended discussion and their significance, see Filler (2004) for a detailed summary; see esp., pp. 1541–9.

4 For a representative example, see Santiago and Olkon (2005).

5 As many scholars have noted. See esp. Kincaid (1998).

6 See Sipe (2004) for a full discussion of Symonds's representations of women in the city.

7 It is perhaps inevitable that the trends of what has been called "chick lit" – *Sex and the City*, *Bridget Jones' Diary*, etc. – would combine with the second-generation immigrant tale to celebrate the second-generation immigrant's enjoyment of sexual and personal possibility in the setting of the consumer city. The ill-fated novel, *How Opal Mehta Got Kissed, Got Wild and Got a Life* (2005), by the young undergraduate Kaavya Viswanathan, was revealed to have been plagiarized from multiple sources, including various chick lit novels and a "young-adult" novel about growing up Indian-American. (It was subsequently withdrawn by the press.) What is notable is that these two genres presented a natural and easy path of generic hybridization, in part because young women have traditionally been portrayed as changing their home-grown mores and finding freedom in the city, and the conflict between home control and urban freedom can be easily dovetailed with the generational conflicts between immigrant and US (consumer, urban) culture, often figured as conflicts between modernity and premodern values.

8 The geopolitics of historiography from Europe to the Americas

Santa Arias

As Geography without History seemeth a carkasse without motion, so History without Geography wandereth as a vagrant without a certain habitation.

(John Smith 1624: 169)

In the study of the (early) modern history of "the invention of America" (O'Gorman 1958) in the European imaginary, the complex relationship and dependency between history and geography is implied and manifested in multiplicity of discourses that encompass political treatises, historical accounts, legal documents, biographies, and memoirs.[1] As John Smith underscored in 1624, history could only be viewed as the dynamic element that brings life to the territory, while space provides "habitation" to what he considered the "vagrant" of history, as it seemed without its stage. In contemporary critical theory, historical space is no longer viewed within the Cartesian and absolute framework that defined most historical discourses of the Enlightenment. More than a scenario of events or the body in need of habitation, space clearly reconnects with history, politics, and culture as we rethink identity, ideas of empire, and society in eighteenth-century intellectual culture. Here I seek to explore the representation and perception of space in historical writing and to see how the study of space or spatiality helps track mobility, flux, and boundary redefinitions found in historical accounts and represented in the cartography of the Americas during the Age of the Enlightenment. When it comes to the understanding of the cultural meanings and the geopolitical processes that shaped this period, regardless of how modern historians view and use space in the twenty-first century, history and geography are interrelated fields that support the understanding of societies. Moreover, we can find how these intellectuals brought forward their cultural milieu imbued with ideas of progress, virtue, and reason to impose Western civilization on the colonized territories. Mapping became emblematic of encyclopedic knowledge as it gave order and coherence to the complexity of the physicality and humanity of the world.[2]

The principal issue that I will interrogate concerns the role that geographical knowledge played in the histories of America by the Scottish historian William Robertson (1721–93) and the Spanish Official Royal Cosmographer Juan Bautista Muñoz (1745–99). Their work is representative of the historiography of this so-called "Age of Reason," when much of philosophical reflection on emancipation and equality still worked under the shadow of imperialism and paradoxically justified domination. Beyond expeditionary accounts, the Americas were the subject of numerous political and philosophical histories by intellectuals who never crossed the Atlantic; however, these discourses inscribed the shape of the continent, its landscapes, environments, and indigenous populations in the European consciousness. They worked toward a more nuanced perspective by placing empirical information within the spectacle of current philosophical debates, reforms, and territorial imperatives. What role did geography play in historical writing and in the philosophical questions such as the nature of Amerindian societies and the rights to rule these territories? Why and how was the description of the land crucial for national histories and sovereignty? More importantly, how did the historiography of the Enlightenment provide a model to understand the role of space in the understanding of history?

I approach these questions not as a historian or geographer, but as a scholar interested in how disciplinary boundaries became irrelevant when the issues at stake concern issues of identity, changing social contexts, and meaning. In the eighteenth century, these issues were deeply connected to territorial control and resistance. While I try to explain the place of geography in history, I will look at one instance within two different national traditions. Within the context of the European Enlightenment, comparative work certainly sharpens up our views of the cultural production of individual empires and how intellectually interconnected they were. While empirical representations of American territories, in the immense body of evidence that supported imperial aims in historical writing, influenced international polarities and helped the reconstruction of particular national identities, they also manifest intertextual relations and dependency that erased intellectual and physical borders.

Understanding the divide

When we look at the basic disciplinary and discursive differences between history and geography, it is obvious that the divide is not that wide. The existence of the subfield of historical geography, and the fact that history has had a constant preoccupation with geography, demonstrate not only common historical origins of the disciplines, but the sustained influence of one on the other. As we see in the chapters in this volume, critical geographers put spatial production on the agenda of the humanities and social sciences by facilitating the understanding of space as a product of ideologies, contingent social forces, and history itself. In historical scholarship, beyond the limiting views of geography as a container or stage of events, considerations of space and spatiality have helped us visualize the impact of power structures, culture, and the market on nature, rural, and urban human spaces.

The materialist view of space during the twentieth century served as the theoretical axis to the history of spatial representations. Well-known critical geographers such as David Harvey, Derek Gregory, and Edward Soja, among others, contribute to the understanding of the social production of space, its subjectivity and historical ontology. David Harvey has underscored how geography is grounded in history as social practices and processes create spaces (1984). Geographical knowledge can no longer be considered a container of life and nature; this absolutist perspective of space has been replaced by a relational one, where space depends on and it is intertwined with historical events, responds to political paradigm shifts, economic changes, and the cultural transformation of societies. Essential to interventions on the role of historical understanding in geographical thought, Derek Gregory's critical geography called attention to Foucault's work on space, power, and knowledge to produce what he calls "the world as exhibition." Similarly, Edward Soja, in his critique of historicism, defined these processes as a "socio spatial dialectic"; for him, "space is a product of social translation, transformation, and experience" (1989: 80). His insertion of space in social theory makes historical and geographical materialism the medium to interpret and understand the relation between spatiality and being human. As he puts it: "New possibilities are being generated from this creative commingling, possibilities for a simultaneously historical and geographical materialism; a triple dialectic of space, time, and social being; a transformative re-theorization of the relations between history, geography and modernity" (1989: 12).

Critical geographers have continued to erase disciplinary boundaries in order to reflect on the spatiality of social life. Their work could have not been possible without the influence of Michel Foucault and Henri Lefevbre, whose thought, as Edward Soja explains in this volume, resonates in the reflection of past and present spaces of history and material analysis. Lefebvre marked the anxieties of the relationship between space and history in contemporary critical geography. For him, space was understandably political, ideological, and a strategic practice that "propounds and presupposes it [space], in a dialectical interaction" (1991: 38). In *The Production of Space,* he drew attention to how every social space has its history and has become "an object of struggle itself" (Elden 2004: 183). On the other hand, Foucault, going beyond his often quoted statements vindicating geography in the interview in *Hérodote* (1984), underscores that space has to be related not only to the discussion of knowledge and power, but also to debates about war, medicine, and science in general. Foucault was interested in the geographers' understanding of power and the meaning of spatial knowledge as a form of science.[3]

The work of critical geographers presents a poststructuralist perspective of space that is central to the understanding of history and its genealogy. As stated by Fredric Jameson, "Why should landscape be any less dramatic than the event" (Jameson 1991: 364). In this line of thought, if landscape – and its representation – must be considered an event, then the analysis on space and spatiality must be historical; after all, space and history are imbued in each other. It is not that geography is key to history, as many intellectuals underscored at different moments during the twentieth century,[4] any real understanding of societal history must consider contingencies that space and place bring to any analysis.

As historians well understand, their field represents more than simple periodization; historical discourses are simultaneously political, social, cultural, geographical, and rhetorical. If we take a closer look at the transformation of the academic curriculum, with critical thinking as a major teaching objective in the humanities and social sciences, a recent tendency has been to work thematically with history, geography, and culture as interconnecting branches. History, with its many new subdisciplines, has always been geographical and is delving more and more into critical spatial relations, as the recent emphasis on transnational studies demonstrates. As expressed recently in a dialog by historians ('AHR conversation: on transnational studies' 2006), the shift in the discipline is more than a focus in comparative history. Scholars are situating historical work within larger frameworks (Seed on 'AHR conversation' 2006) and beyond the study of historical periods or regions; transnational history concerns movement and circulation as a method that is defining the field (Hofmeyr on 'AHR conversation' 2006).

Beyond the often-taught world history survey, or national and regional histories, research and teaching in history calls for a strong consciousness of space and place where global, immigration, diasporic, and many other forms of history-making open the world of possibilities. New approaches show the influence of current debates, such as globalization, world-systems theory, border crossing, uneven development, and (post)coloniality, among others. It is interesting how the emergence of Atlantic or Pacific histories (representing the history of oceans that used to be considered a historical void) have become subfields that help interpret the past while transcending national histories. Shifts in historical research have given space and spatiality a privileged position for the study of social and political interrelations and networks that produce history.

History and geography during the Enlightenment

As seen by the extensive production and circulation of histories and accounts of the Americas, Asia, or Africa, colonial knowledge that reflected imperial aspirations focused on the depiction and labeling of conquered and contested spaces. During the late eighteenth century, intellectuals from modern states such as France, England, and Spain played a key role in the expansion of Europe's human, economic, and scientific understanding of the American territories. The reception of geographical and historical knowledge shaped significant "Enlightenment sensitivities toward America" (Withers 2007: 158) that gave prominent roles to natural, civil, and geographical accounts in the national discourses of the history of the era. While Spain defended its colonial undertakings and pioneering contributions to geography, England and France made claims to territories with historical accounts that also placed geography at the core of progress – supporting commerce, navigation, and scientific achievements as legitimating evidence of supremacy. We must remember that during this period American territories (British and Spanish) were negotiating and slowly securing their independence. Regardless of the shifting historical and ideological frameworks of these nations, their geopolitical discourses were unveiled within an absolutist idea of space. Within this perspective, education was supported on principles of the Enlightenment and defense of different European structures of power and identities.

The late eighteenth century, a crucial period for the development of Western geography, is considered one of the busiest periods of well-documented expeditionary travel, such as those by James Cook (1769–80), Louis Antoine de Bougainville (1766–9), the Count of La Pérouse (1785–8), and Alejandro Malaspina (1789–94), among others. These navigators and scientists contributed enormously to the empirical knowledge of the territories in the Atlantic and Pacific oceans, to the culture and discourse of science, and, within those, to accounts of human difference. Space translated into race in the philosophical undertakings of intellectuals of the period, demonstrating that interrelations, movement, and flows of societies opened up broader analytical possibilities for understanding humanity and geopolitical processes. While intellectuals were justifying empire, they were also engaged in transcending it.[5] Mapping became an instrument of empire; it linked history and political events to place and "supported a growing sense of sovereign political bodies" (Black 2000: 12). Greater accuracy was obtained with the problem of longitude resolved and a larger demand for history books that provided greater geographical detail and accuracy: "Book readers sought a history informed by precise cartography and a cartography that was historically accurate" (Black 2000: 8)

At the center of late eighteenth-century historical discourses about unknown – and desired – territories was Spain, a political power that became the subject of severe attacks that resuscitated sixteenth-century wounds initiated by the Protestants.[6] The renowned *leyenda negra* (Black Legend) re-emerged and fed much of the critique of Spanish violent colonial practices over the indigenous populations, in spite of the persistent claims that the main objective of the conquest was to bring the Catholic faith to the Americas. Critics of Spain pointed as evidence to the writings of the sixteenth-century Dominican Bartolomé de las Casas, renowned for his defense of the Amerindian populations at the time of contact. His *Short Account of the Destruction of the Indies* (1552) was projected as evidence of the greed and atrocities committed in the Spanish regions in the name of Christianity. This text, a political treatise that was used as a geopolitical tool against Spain, presented a geography of violence that scandalized Europe over the exercise of Spanish power in colonial territories. Las Casas collected and edited eyewitness testimonies of physical and human destruction in every island and region of the continent. His rhetoric was grounded in spatial interrelations where the methods of aggression in one place influenced other spaces as colonialism extended throughout in the hemisphere. Las Casas's manuscripts circulated widely during the eighteenth century, generating a polemical debate about the consequences of empire. Voltaire, Cornelius de Pauw, Guillaume abbè Raynal, William Robertson, and Juan Bautista Muñoz all expressed strong opinions on Las Casas.[7]

More than an assessment of the Spanish, at the core of late eighteenth-century histories about the Americas was the need for validation of control of territories that sustained the interest of the rest of Europe. Therefore, the revival of the Black Legend served to demonize Spain and, as it has been recently argued, supported the construction of whiteness that gave rise to modernity and vested dark symbolic power in anything Hispanic in the Anglo-American imagination (DeGuzmán 2005). Within that modernity, we must account for trade and expansion, as intellectuals rethought Buffonian ideas about the place of indigenous societies in nature.

Late eighteenth-century intellectuals framed what Antonello Gerbi (1993) called the "dispute of the New World" with geography at the center. They elaborated a pervasive description of space in the Americas that encompassed a deterministic approach to human diversity. These views emphasized the influence of the environment on indigenous populations and long-term effects over European settlers. Montesquieu, Buffon, Raynal, and De Pauw, among others, recycled classical and early modern thought from Hippocrates to Jean Bodin. For them, the physical environment influenced mental and physical states, as well as societal development. In spite of of how contemporary scholars understand the backwardness of environmental determinism as a measure to understand societies, in the late eighteenth century it became a major subtheme in geographical discourse and made geography more relevant in philosophical inquiries and in historiography. Most of the historical accounts of the colonial period, from the sixteenth-century official historian Gonzalo Fernández de Oviedo's *Historia general y natural de las Indias* (1535) to the eighteenth-century Chilean Creole, Juan Ignacio de Molina's *Compendio della storia geographica, naturale e civile del Regno de Chili* (1776), included extensive geographical narratives that articulated this conundrum while they tried to affirm or deny the humanity of Amerindian societies. These historical accounts, beyond locating regions in their longitudes and latitudes, also expanded and interpreted knowledge on topography, hydrography, climate, populations, flora, and fauna. Furthermore, in the eighteenth century, geographical knowledge demonstrated that the domination of nature and indigenous societies was essential in the expansion of political borders and, as we consider the Creole response to European histories, it facilitated an affirmation of identity embedded in their land. In the case of William Robertson and Juan Bautista Muñoz, while they represented and defended their nations, they also presented a critical and comparative perspective based on rationalism and primary sources that set them apart from other intellectuals of the period. They were not writing in a vacuum.

Writing the Americas

While Diderot and d'Alembert's encyclopedic project unfolded (1751–65) with geography framed within history, more attention was to be placed on the Americas.[8] The decade of the 1770s was known for the centrality of this region in new critical European philosophical and political histories. Hence, late eighteenth-century historians, such as Robertson and Muñoz, set out to reread early colonial accounts and find in old forgotten archives more sources to complete and give new meaning to the early years of contact. Textual evidence validated new reforms such as the open trade put in place by Charles III, defined social hierarchies, and helped enforce control over the territories. For them it was clear that the benefits of European domination (agriculture, commerce, religion) outweighed the pain of the wars of conquest. In the case of the Scottish historian, impartiality seemed to rule his rhetoric when he set forth the ideological agendas of the period that were influenced by the new ideas of progress, power, and by the time he published his *History of America*, the loss of British colonies and emergence of a new American identity.[9]

In order to interpret the impact of the "new discoveries" within the wider frame of European history, William Robertson led European historians when he published in London (with an immediate reprint in Dublin) his ambitious *History of America* (1777).[10] Using a range of historical registers and primary sources carefully listed after his preface as evidence of his reliability as a historian, he proposed to analyze the impact of the early "achievements and institutions" (1851: p. iv) by the Spanish, Portuguese, and British, and the nature of the physical and human space of the continent. Book 4 is perhaps the key part of the project where he takes on previous historians to find a middle ground in the study of the nature of Amerindians. He compares them to other societies that were believed to be either more primitive or the more advanced European civilizations from antiquity. Both Robertson and Muñoz recognized, expanded, and revised the ideas of Buffon, Raynal, and De Pauw. While Muñoz refuted them fiercely, Robertson, in a very complex argument, supported a modified Buffonian deterministic approach by stating that temperate zones influenced Amerindian societies in a positive manner while in the Tropics they were violent and degenerate. In general, he proposed a four-stage theory of social development based on geography and modes of survival.

The first edition of the *History of America* did not include the British territories or the Portuguese settlement in Brazil. Robertson explained that his history was incomplete; however, he went ahead because "of the present state of the British colonies" and questions about "ancient form of policy and laws" that civil war brought forward (1851: p. i).[11] Here he was referring to the Declaration of Independence that had dissolved British claims to American territories within less than a year of the first printing; Robertson's arguments on British colonial rights and supremacy had no point (Armitage 1995: 69).[12] He continued making revisions for subsequent printings until 1788. After his death, the papers that comprise the histories of Virginia and New England (books 9 and 10) were found and published by his son in 1796 (Humphreys 1969: 25). As David Armitage (1995) stated, no other historical account could compare to Robertson's since Antonio de Herrera's multivolume *Historia general de los hechos de los castellanos* (1601–15). Nevertheless, his historical assessment had an underlying agenda since Robertson believed that the Indians were indeed an inferior race and that European political domination over American spaces was justified and necessary.

After its publication, Robertson's *History of America* became a bestseller and praised for his method and critical insight (Sher 2006). Detailed notes followed the main body of the text reflecting the scope of his research and thought on issues ranging from the history of the classical period to events during his lifetime. In Spain, the Royal Academy of History approved its translation and publication. Nevertheless, this attempt was halted when an outside reviewer pointed to Pedro Rodríguez de Campomanes and members of the Royal Academy that the text "at every turn insulted Spain" and argued that "only Spaniards could write a truly history of the Indies" (Cañizares-Esguerra 2001: 178–9). Robertson, who maintained a relationship with Campomanes, managed to represent the Spanish colonists as greedy and cruel. For him, their major mistake was their lust for gold when they should have been investing in agriculture, trade, and commerce; for Robertson "Their system was not based in toleration and civil liberty" (Phillipson 1997: 62). Robertson's *History* was banned in Spain, Spanish America, and the Philippines by royal decree in 1778. At that point, and with Spain in much need of a corrective history, Charles III appointed Juan Bautista Muñoz to write the historical account of Spain in the Americas. Muñoz accepted the commission, an event surrounded by much debate since the Royal Academy of History held all official responsibilities dealing with national history. Muñoz had to write a new critical history that would re-evaluate Spain's own colonial experience and respond to the negative representation that the *philosophes* had promoted in Europe. Spain's attempt at a corrective history came late in the eighteenth century, after the country had suffered many years of criticism hurting its own national identity and image as a modern European power.

In 1793, the same year that Robertson died, Juan Bautista Muñoz published his *Historia del Nuevo Mundo* to try to save Spain´s colonial past from the harsh critique that was endangering its rule in the Americas. As can be imagined, with a booming book industry in Britain that had a preference for histories, Muñoz's *Historia del Nuevo-Mundo* was translated into English and published in London in 1797. Surprisingly, the preface to the English edition introduces Muñoz as the historian of the Americas who surpassed others for accuracy and a comprehensive design to "dissipate any clouds which hung over the History of the Discovery of the New World" (1797: p. vii). In spite of the several attempts by Robertson to procure as many original sources as possible to compose his history, Muñoz, for his *Historia*, succeeded where Robertson had failed.[13]

Muñoz's *Historia* achieved more than a defense of the virtues of the continent, its nature and environment. The first volume reassessed Spanish history and emphasized Columbus's contributions to geographical knowledge and navigation in the reconnaissance of the western unknown regions. He also intended to write about the conquest of Mexico and Peru, but his second volume was never published.[14] Muñoz used as a major historical source Bartolomé de las Casas's writings on the Columbian enterprise, which defended Columbus while paradoxically offering his aggressive critique of early Spanish colonial practices.

Ironically, to write his history Muñoz followed the model set in William Robertson's *History of America*. Perhaps the Spanish historian was the author of the anonymous review that banned Robertson's text and halted its translation (Cañizares-Esguerra 2001: 181). Ironically, in trying to refute Robertson, who wrote under the shadow of Buffon, Raynal, and De Pauw, Muñoz followed his arguments closely.[15] Nevertheless, he was not attempting emulation but refutation, and highlighting Spain's achievements in science, navigation, and commerce.

Robertson's reflections on geography had three angles: first, in his statement of the centrality of commerce and its dependency in new geographical knowledge; second, in the demonstration of the role of geography to understand the population of indigenous societies; and third, in the philosophical exegesis on the influence of location on behavior. In the first book, he offers a painstaking description of early European navigation where advances in geography and branches of science are linked to the Greek, Roman, Arab, Portuguese, and Spanish contributions to navigation and, consequently, to the civilizing role of commerce. His comments on geography were not as explicit as those by Muñoz, however, in the notes to the text he elaborates the relevance of space, place, and mapping in his reconstruction of history. In a comment on Robertson's work, Major Rennell (a geographer) states:

> Since I understood the subject, I have ever thought, that the best historian is the best geographer; and if historians would direct a proper person, skilled in the principles of geography, to embody (as I may say) their ideas for them, the historian would find himself better served, than by relying on those who may properly be styled *mapmakers*.
>
> (1777: p. cxiii)

Rennell considered Robertson the best historian to take on such an enormous task because of his knowledge of geography.

Mapping the spaces of history

As usual with historians of the Enlightenment, William Robertson and Juan Bautista Muñoz had prefatory maps that complemented their historical accounts and served as rhetorical and legitimating instruments of the "truth" conveyed in the text. In addition, maps made historical accounts more valuable in the flourishing European book market. Beyond the aesthetics of the period, these maps linked and illustrated the relationship between historical inquiry and territorial possessions to national politics and commercial expansion within the unsettled global landscape of their empires. The *History of America* included four maps by Thomas Kitchin (1718–84), the royal hydrographer, who was considered one of the most prolific map-makers of the period.[16] These historical maps were intended to visualize and spatialize post-conquest history with exceptions that pointed to precolonial divisions of the territory or the establishment of new settlements significant in the 1770s. Kitchin's map of South America, included in Robertson's *History*, indicated with an intermittent line European possessions at the time (Figure 8.1). However, in his "Map of the Golf of Mexico" in Hispaniola he incorporated the original Taino divisions of the island as at the time of Columbus's arrival. This map shows the influence of French cartographers whose maps reflected France's long-standing interest in Hispaniola.

The four maps included were busy, as they layered history by locating old and new places and continued uncertainty over the naming of regions. For example, one feature found in much of the cartography of the period was the hesitation at choosing toponyms: "Mexico or New Spain," "Pacific Ocean or South Sea",

Figure 8.1 "Map of South America," by Thomas Kitchin. Courtesy of Robert Manning Strozier Library Special Collections, Florida State University.

"Gulf of Mexico or North Sea", "Venezuela or Caracas". Names changed as regions changed hands or as the Creole intelligentsia intervened in the naming of their own territories.

Beneath the spaces represented on paper we find the social forces behind Europe's civilizing mission. Kitchin's map "Mexico or New Spain, in which the motions of Cortés may be traced" acknowledge in detail physical aspects of the territory that includes the topography, rivers, lakes, coastlines and harbors; as well as human spaces established by indigenous groups, Spanish missionaries or British settlers (Figure 8.2). This map was done with more care and detail than those commissioned to Kitchin for inclusion in Raynal's *Histoire philosophique et politique* (1770). He combined in his depiction French cartography, especially d'Anville's representation of the Americas ("Hemisphere Occidental ou du Nouveau Monde," 1761), with newer revisions done according to instructions and new knowledge of expeditions of the interior. Kitchin's visualization of the territories relied on what was considered trustworthy information and when none was existent, he did not hesitated on labeling it as unknown. In this map, New Mexico borders to the north a "Great Space of Land unknown" that erases from the landscape the indigenous groups populating the Great Plains region. As in much of Robertson's work, the map also shows ambivalence towards the Amerindians. In the case of Spain's possessions, Amerindian place-names are clearly marked intermingling with Spanish settlements. In most European maps of the colonial period, the cartographer controls the location of indigenous settlements. These regions are omitted to leave space to other places where crucial historical events took place or where major colonial institutions were located. In the map "Mexico or New Spain" the promised route of Cortés is buried, and the viewer can barely distinguish important places of his itinerary such as his first establishment of Veracruz, or the indigenous Zempoalla (Cempoala), or Tlaxcala. The "environs of Mexico" referring to Technotilan and its surroundings areas are highlighted in the insert. It provides further details on a smaller scale of the Aztec region around the "lake of Mexico" where the historical *calzadas*, water sources and urban design are clearly marked.

Muñoz intended to include three maps, one of them representing the Caribbean region, described similarly to Kitchin's Map of the Gulf of Mexico that was included in Robertson's. Beyond the tactical inclusion of maps to show the scenario of events, Muñoz embedded the history of the region within a reflection of the world commercial system. For Muñoz, Columbus, who represented Spanish interests, was the key figure in the emergence of modernity. He established the existence of a new continent and redefined European seapower. For Muñoz and Robertson, the Atlantic became the space of global politics, with Spain and England respectively leading in the colonization (as synonym of progress) of North and South America.

In Muñoz's *Historia del Nuevo Mundo*, the artist Tomás López Enguídanos (d. 1814) drew the prominent map "Mapa del Nuevo-Mundo" with its insert of Hispaniola (Figure 8.3).[17] In the hemispheric map of the original publication, the cartographer followed a different approach to Kitchin. López's map was clean, with no sign of frontier lines, layering colonial history with its precolonial and colonial place names but pointing to new discoveries by correcting coastlines that show his knowledge of new voyages such as those of the northwest region

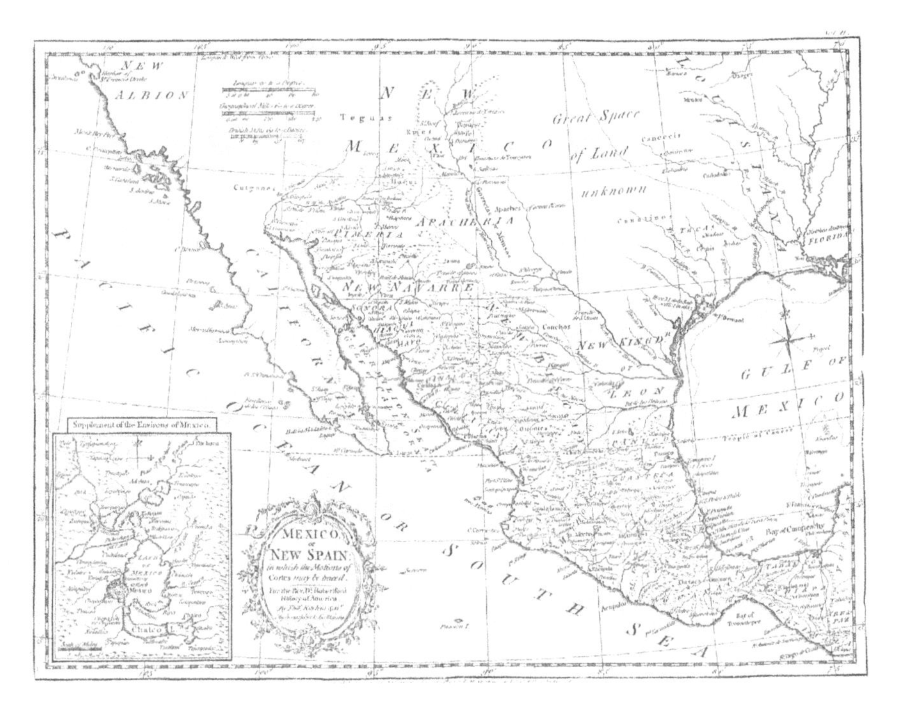

Figure 8.2 "Mexico of New Spain, in which the Motions of Cortés may be traced," by Thomas Kitchin. Courtesy of Robert Manning Strozier Library Special Collections, Florida State University.

of North America. Muñoz explains "These are the western discoveries of the Spaniards. This is the New World worthy of the name not only on account of its been unknown to the ancients, but even for the new things which it affords and produces in physics and morals" (1797: 70). If Robertson placed commerce and navigation within the broader scope of universal history, in Muñoz's text, Spain's colonial accomplishments led Europe into that same modernity defined by navigation and what the Spaniards contributed to knowledge: "In consequence of these discoveries, the southern part of our globe was circumnavigated, and its true form brought to light, together with a knowledge of its principal parts and products, and what was still greater, the sphere of our ideas was enlarged" (1797: 70).

Both texts and maps attempted to secure physical spaces; while Muñoz legitimated Spain's political power, Robertson's placed European rights within its civilizing force. Space as territory or as "produced, perceived and lived realms" (Lefebvre 1991: 40) is deeply linked to identity formation and the subject's mobility. For both of them, Columbus, Cortés, and Pizarro moved through territory, appropriating it in the name of religion and civilization. For Enlightenment

Figure 8.3 "Mapa del Nuevo-mundo," by Thomás López Enguidanos. Courtesy of Robert Manning Strozier Library Special Collections, Florida State University.

intellectuals, they provided an insight into nature, the world, and human diversity. They led the first colonists to occupy, organize, and control space.

Conclusion

In the study of historiography, it has been recognized for quite some time its rhetorical force as historical narratives incorporate different discourses (i.e., legal, religious, autobiographical) and combine tropes to construct their plot and persuade as necessary (White 1978, 1987). In the eighteenth century, maps and charts included in histories played a major role as another form of persuasive expression and consistent emphasis on the importance of territory. As J.B. Harley explained, in the interpretation of maps, we have to move beyond "the assumed link between reality and representation that has dominated cartographic thinking" (1992: 232) and find gaps and silences that cannot be explained within positivist epistemology. Indeed, cartography provided the material specificity needed for history-making, offering an interpretation of territory as an essential commodity and representing a subject of contention among political powers.

What is the proper place of geography in the study of history? The historiography of the Enlightenment concerning the Americas presents one of the best examples of the production and use of geographical knowledge to interpret history. It is clear from the study of Robertson and Muñoz how both disciplines are bonded together and as they produced knowledge about a place, they express power within a wider context of inquiry about human diversity and debate about territorial claims. For William Robertson and Juan Bautista Muñoz, writing the histories of the Americas was not a project of description of Europe's possessions, but a reflection of the colonizing project of England and Spain that revealed that geography was too pervasive and important to the understanding of difference and progress in science and commerce. They spatialized imperial aims by providing a geographical narrative that pointed to the layers of history of their maps. In these accounts, they challenged historical assumptions and repositioned their nations with all the authority that historical discourse could provide. Their accounts presented space as a place of interrelations in time and space, thus redefining Enlightenment ideology and modernity.

Notes

1 Edmundo O'Gorman (1958) explains how the ideas of the Americas disseminated in Europe were constructed according to particular loci of enunciation that did not correspond to the actual reality of Amerindian societies at the time of contact.
2 On the subject of the relevance of geography and cartography during the Enlightenment, the work of David Livingstone (1992, 2003), Matthew Edney (1997), and Charles Withers (2007) are paramount.
3 For more on Foucault's contribution to geography, see Elden and Crampton (2007).
4 See Alan Baker's assessment on the interrelations of geography and history in the twentieth history (2003: 16–24).
5 In a recent dialog between historians, "transnational history" was defined as a field that "implies a comparison between the contemporary movement of groups, goods, technology, or people across national borders and the transit of similar or related objects in an earlier time" (Seed in 'AHR conversation' 2006: 1443). It should not to be confused with "global history" or the long-standing subfield of "world history."
6 There is an extensive literature on this topic. Detailed accounts on the "Black Legend" include the classical studies by Julián Juderías (1914) and Rómulo Carbia (1943); in American history and cultural studies by Charles Gibson (1971), William S. Maltby (1971), and, more recently, by María DeGuzmán (2005).
7 For Muñoz, the Dominican's writings offered detailed information about geography, culture, and the environment of the Americas that, regardless of his accounts of atrocities, supported imperial claims at a critical moment when Spain was losing its territories (Arias 2007).
8 According to the encyclopedists "Geography should be placed within three different period headings" and with clear subdisciplines "Natural, Historical, Civil and Political, Sacred, Ecclesiastical and Physical" (quoted in Withers 2007: 174).
9 On the role of impartiality in Robertson's *History of America*, see Smitten (1985).
10 Robertson's *History of America* circulated widely. Its many editions and reprints included revisions by the author. He also became the best-paid Scottish author of his period in the emerging book market in Britain (Sher 2006: 214).

11 Humphreys interpreted his statement by saying "Robertson preferred to postpone his discussion of the British colonies until he should know what was going to happen to them. He would wait, he said, 'with the solicitude of a good citizen, until the fervent subside, and regular government be reestablished, and then I shall return to do this part of my work'" (1969: 24).
12 On Robertson's conservationism and his inability to finish the historical portion on British America, see Smitten (1990).
13 Robertson could not travel to Spain and his intermediaries did not have access to the Archive of Simancas. Most of the materials that he was able to compile were acquired by purchase (Black 2000: 120).
14 An incomplete second volume is part of the Obadiah Rich collections housed at the New York Public Library.
15 As a sign of how closely he modeled his own history after Robertson, his official proposal presented to the Council of Indies, in order to get approval for the publication, was entitled *Historia de América*, a direct translation of Robertson's book.
16 Kitchin had partnerships with some of the most renowned cartographers of the period in England, among them, Emanuel Bowen with whom he published in 1755, "The Large English Atlas".
17 The English translation did not include the map, but only the insert and a portrait of Columbus.

9 "To see a world in a grain of sand"

Space and place on an ethnographical journey in Colombia

Margarita Serje

Several years ago I took part in a utopic experience inspired by the Sierra Nevada de Santa Marta in Colombia. This massif, insulated from the Andes, hosts almost the entire spectrum of tropical American ecosystems, as well as a great number of endemic species. By the time of the Conquest, the Sierra was inhabited by a sizeable population that built settlements upon an astounding stone infrastructure. This society, known as the Tairona, is considered along with the Muisca to be one of the great precolonial civilizations in Colombia. Today, four indigenous groups, descendants of these builders of stone, inhabit the highlands of the Sierra. By virtue of the intriguing world of esoteric knowledge with which "the magic of ethnography" (Malinowski 1961) has endowed them, the peoples of Sierra Nevada appear as the "guardians of the world" (Davis 2004: 38) whose "voices of ancient wisdom rise to save the planet from pollution" (Perera 1998: M2).

In the early 1970s, the vestiges of one of the ancient Tairona cities were discovered in the slopes of the Sierra facing the Caribbean. The Colombian Anthropological Institute (ICAN) set up a research station in the site known as *Ciudad Perdida* (Lost City). I was part of the team of researchers who arrived in 1979 to work in this area, shrouded in mountain rainforests. The beauty and profusion of these forests led our group to propose that recovering the archaeological sites should not only be aimed at studying the social organization of the Tairona, but also their agency in environmental management. It was important to consider not only the archaeological remains, but also the forest itself, insofar as its history is inextricably linked to the indigenous way of life, both pre-Hispanic and present.

We asked ourselves how it was possible that after intensive occupation lasting close to 500 years these forests continued renewing themselves, whereas as a result of recent occupation in less than 30 years, the area was already covered by pastures, and erosion was irreversible. By then, the marijuana boom was in full swing. The *Santa Marta Golden* (as the marijuana harvested in the Sierra Nevada was known in the illegal market) advanced like a shadow that decimated in its wake thousands of hectares of forest. With it, contingents of peasant-settlers penetrated upstream, pushing indigenous communities to higher ground. We set out to establish a permanent presence on the ground, where the challenge was to coexist with the neighbors, settlers or *indígenas*. For us, our work was no longer just about studying archaeological ruins, but about making new knowledge. Thus, studying "the past for the future" became one of our main objectives. We delimited an area that was to become the "Historical and Natural Park of the Buritaca River", where it would be possible to fulfill two major objectives. The first was to protect, at least in this zone, the integrity of indigenous territory in the face of violent onslaught by marijuana growers; second, to protect the archaeological sites and the forest, both equally valuable as evidence of ancient settlement and as contemporary indigenous landscape and way of life. To achieve our goals, we built "stations" in strategic sites. We were aware of the fact that our physical presence would illustrate the type of relationship we intended to establish with our surroundings. Thus, we decided that our dwellings should reproduce in their architecture and landscaping the models provided by the indigenous settlements (both current and ancient) in terms of materials, water management, vegetation and topographic implantation. Indigenous authorities came by often to find out what we were doing with the sites where the "Ancients became stone" and they explicitly asked us not to dig. Archaeological remains are for them places where the "mothers" of ritual knowledge and healing are stored. We agreed to stop archaeological excavations, devoting ourselves instead to studying what had already been looted and to regional prospecting.

In our zeal to protect the area, we also began to get acquainted with *campesino* settlers. We met people who came from the most violent areas of the country. The "old ones" had arrived during the decade known as La Violencia (1948–58)[1] attracted by the "gold rush" of looting. Back then, they told us, gold artifacts came out in heaps. Behind the looters, peasant-settlers started building homes on the Tairona stone infrastructure that appears to cover the entire northwestern corner of the Sierra. As the *guaquería* (tomb-raiding) was in full swing, marijuana made its appearance. Attracted by this new boom, a second contingent of settlers arrived. When we set up the Historical Park, this boom was at its peak. Management of the project changed hands to an NGO created in order to streamline its financing. This scheme was the object of a heated public debate, which I will not discuss here, that concluded in the mid-1980s with the closing of the project and the expulsion of all of us who worked there. By then, marijuana was already beginning its decline and violence and poverty were evident in the peasant region.

The project's closing, however, did not to put a halt to the commitment that our group felt to its friends and neighbors, indigenous as well as settlers. We resolved to create a foundation, modeled after the one that was being dissolved, with the capacity to receive financing from both the public and the private sectors, in order to continue with our endeavor. Thus, in 1986 we created the Fundación Pro Sierra Nevada. One of its first goals was to set up a "community support center" in order to work directly with the peasant population. The premise was that by strengthening this group of people and by proposing a form of habitation that was in accordance with the principles of sustainability learned from indigenous and Tairona settlements, it would be possible to secure their welfare while at the same time reducing their pressure on indigenous territory and its forests. We thus began to work with settlers along the usual lines: infrastructure, health, small productive projects, training community leaders and providing support and advice for interactions with government bodies.

Then the guerrillas arrived. At first they only made sporadic appearances, but finally they stayed. They began by holding meetings with the community, visiting ranches, planning collective work, and settling everyday conflicts. They warned that those who behaved dubiously would be severely punished. When I asked one of our peasant neighbors how she saw the intervention of the guerrillas, she described it in a single phrase: "the guerrillas do the same thing you do, only with weapons." I was still disconcerted by those words when, a few days later, a mule train caught up with us along the way. There was no mule driver. Each mule had a bloody corpse tied to its back: all of them were peasant neighbors from the area. As the guerrillas admitted a few days later in a community meeting, the idea was to do some "social cleansing" in the region, and for this purpose they had executed all of those they considered "problematic".

The thought that we did the same as the guerrillas, only without weapons, put an end to my idyllic view of the Park and confronted me with a series of questions that have haunted me ever since. In trying to decipher this woman's assertion, I was forced to see with different eyes what it means to try to promote order in the life of a population and its landscapes. I began to realize that our work group had transformed this place into a project dependent on our way of conceptualizing the Sierra as landscape and as the setting for our utopian vision.

My objective in this chapter is to illustrate, through our experience, beginning with an archaeological site and leading to the invention of a "planning region," some of anthropology's main concerns for space and spatiality. More than conducting an exhaustive review of the major anthropological approaches to the study of space and landscape and debates on methods and theory,[2] I seek to discuss and to illustrate a series of key issues. I am interested in showing the production of anthropological knowledge as a "spatial practice,"[3] i.e., a strategy through which a spatial reality is created just as the social conditions of its creation are veiled.[4] In order to do this, I will focus on the practice of contextualization, which is key for the production of anthropological knowledge, and in which a discursive field is created as a necessary condition for, in Trouillot's (2003) terms, a geography of the imagination that goes hand in hand with a geography of management.

Here I also propose an ethnography that seeks to travel the paths, both concrete and metaphorical, taken by our group, through which new connections were made between the Sierra and the world. Apart from sketching the itinerary of the invention of this place as a cultural story, I intend to contribute to the visualization and legitimization of territorial notions developed by its inhabitants. From this standpoint I intend to make a contribution to the ethnography of this place, to "the ways in which citizens of the earth constitute their landscapes and take themselves to be connected with them . . . how men and women dwell" (Basso

1996: 54). Although this type of work is not abundant (Hirsch 1995; Feld and Basso 1996), anthropology has always concerned itself with the "aura" of a place, the way in which geography is experienced and endowed with meaning. Ethnography, by delving into the fine grain of everyday life, has tried to "see a world in a grain of sand" in William Blake's famous line, or as Basso puts it, "in a few grains of carefully interpreted sand" (1996: 57), understanding inhabited places as places filled with social and cultural worlds.

The field

Perhaps the first lesson our Historic Park team[5] learned had to do with the instability of the concept of "field." Fieldwork, which constitutes one of the distinctive pillars of the discipline, historically took place through a series of practices that entail an experience of *dépaysement:* of displacement and estrangement, embedded in a certain way of understanding the practice of travel. In this tradition, empathy for worlds located in the confines of the modern is intertwined with a particular way of narrating them (Rabinow 1977; Fabian 1983; Clifford 1988, 1997; Gupta and Fergusson 1997; Trouillot 2003). The practice of the field locates and circumscribes space. Archaeological as well as anthropological analysis took the isomorphism between space, time, and culture as certain (Fabian 1983; Gupta and Fergusson 1992a, 1992b; Thomas 1996), insofar as any understanding of social processes was anchored in the description of everyday life at a locality (a village, an island, a "longhouse") identified with the particular group that inhabits it: "Ethnography thus reflects the circumstantial encounter of the voluntarily displaced anthropologist and the involuntarily localized 'other' " (Appadurai 1988: 16). As pointed out by Trouillot, "increasingly, anthropology's object of observation turned out to be defined primarily as a locality" where there is a tendency to "conceive places as best as locales, and as worst as localities, rather than as locations", in the sense that "its situatedness as locations remains vague" (Trouillot 2003: 122–3).[6] The historical circumstances of the Sierra forced our team to *situate* the field we worked in, expressing the epistemological crisis that began to take form in the discipline.

Our field, which was at first an archaeological site, was temporally destabilizing. When artifacts from the Castilian occupation began to appear, we realized that the "occupation stratum" did not speak so much of the synchronic reality of a pure Amerindian culture, but rather of a social process marked by a colonial relationship. Later, when trying to determine the site's limits, we found that it was not possible to establish a clear perimeter for the cluster of lithic structures: the density of terraces and contention walls decreases along the trails and increases again until it conforms another agglomeration, making it impossible to establish clear limits between them: "The site" had to be expanded to cover the spatial continuum evidenced by the lithic infrastructure. The "field" was again transformed when we introduced the concept of landscape and indigenous historical culture as relevant for archaeological interpretation. All of this forced us to leave aside a defined object of study as a set of traits and objects in order to visualize it as a complex network of social relationships where the field appeared more as an open realm, connected in several ways to the "external" world. In this change, our interaction with the indigenous communities had many pragmatic aspects, including the exchange of products and knowledge.

The "field," then, ceased to be a discrete place – like an island in the middle of the jungle in which one arrived by helicopter – separated from our "real" lives *à la* Malinowski. It was transformed into a territory in which our lives were involved in three ways. First, we embraced the personal nature of our relationships with our neighbors, both indigenous and peasants. Second, our intention to turn the Park itself into a model of habitation for the region forced us to establish an explicit exchange of knowledge, which introduced nuances to the researcher–informant relationship that characterized ethnographic practice. And third, we were under permanent threat because we created a protected area where we effectively prevented access to chainsaws and marijuana planters. All of these factors gave a visceral twist to the meaning this project had. Santa Marta and Bogota ceased to be places where we could comfortably be "outside" the field, and became another battlefield for our endeavor. By the time the Historical Park was closed, it was already a physical entity: ICAN received a "site" that was no longer limited to the Lost City, but rather an area in the Buritaca basin delimited by a series of stations that were both archaeological sites and centers of interaction with our neighbors in the area.

The Fundación Pro Sierra continued with the station program as a model for the region's future. This model represented a *spatial order:* its aim was to recover the topographic patterns, the logic of the lithic infrastructure, and the use of materials, water and vegetation in order to recreate a landscape that reflected the historical settlement of the Sierra. The starting point for this project was understanding the "rational order in remarkably visual aesthetic terms . . . an efficient, rationally organized city [is conceived as] a city that look[s] regimented and orderly in a geometrical sense" (Scott 1998: 4). The visual and geometric order of this "model of habitation" grew out of the fusion of Tairona stonework and Kogui architecture.

The second major objective of the Fundación Pro Sierra Nevada was to call upon the state to pay attention to the Sierra Nevada, showing that it was possible to propose an adequate intervention model for this region. From the state's point of view, the Sierra existed as a marginal area, distant from administrative centers

> this fact has precluded the existence of integral policies and guidelines for the management of the Sierra Nevada as a region, since this task is handled from three administrative centers that all look to the coast. The foundation defines itself by taking the Massif as a territorial unit comprising the area topographically conformed by the Sierra and its piedmont, easily identifiable thanks to its insular nature
>
> (Fundación Pro Sierra Nevada, 1987)

este hecho ha impedido que existan políticas y lineamientos integrales para el manejo de la Sierra Nevada como región, pues la gestión se maneja desde tres centros que miran hacia la costa. La fundación se define entonces tomando el macizo como una unidad territorial que comprende el área que topográficamente conforma la serranía y sus estribaciones, fácilmente identificable debido a su carácter insular

We endeavored to make the massif visible, transforming it into a legible reality in the eyes of the administrations and showing its strategic importance to the nation:

in the Sierra Nevada, the country has its most important ethnic, archaeological, ecological and cultural roots. Nevertheless, the sierra is a fragile environment and its future is not clear

(Mayr* 1984b: 7)

el país tiene en la Sierra Nevada, sus más importantes bienes étnicos, arqueológicos, ecológicos y culturales. Empero, la sierra es un medio frágil y su futuro no está claro

The aim was to make the Sierra more visible, by proposing a specific interpretation of its reality.

This way of situating the Sierra was expressed in the Fundación Pro Sierra Nevada's "creation discourse", rendered in the book *La Sierra Nevada de Santa Marta* (Mayr 1984a) which presents powerful visual images of the region. The book is centered on a series of spectacular photographs that introduce the Sierra Nevada through the primordial nature of a fog-shrouded landscape. The photographs of the snowcapped peaks and the forest covered mountainsides are fused with the opening words of the Kogui creation myth:

First there was the sea. All was dark. There was no sun, no moon, no people, no animals, and no plants. Only the sea was everywhere. The sea was the mother. . . . The mother wasn't a person or anything, it was no thing. She was the spirit of what was to come and she was thought and memory.

(Reichel 1985: 17)[7]

Thus the Sierra/Mother is brought to life in this book through the dramatic nature scenes that appear as setting for the Tairona and the Kogui cultures.

The Fundación Pro Sierra also creates and disseminates a map of the Sierra, keeping in mind that

the Foundation has as its main task to be an advocate for the conservation, protection, investigation and integral development of the geographic, cultural and social complex of the Sierra Nevada massif

(Fundación Pro Sierra Nevada 1987)

la Fundación tiene como tarea principal propender por la conservación, protección, investigación y desarrollo integral del complejo geográfico cultural y social del macizo Sierra Nevada

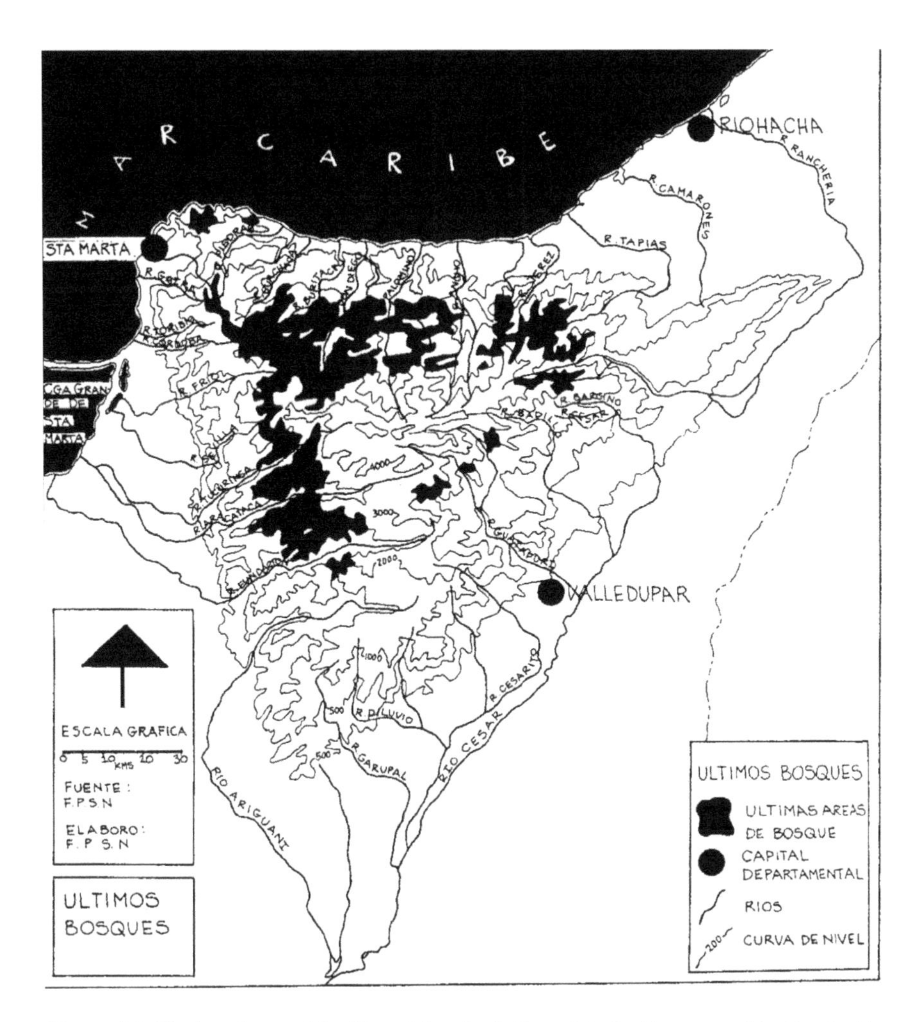

Figure 9.1 The last forest of the Sierra Nevada de Santa María. Courtesy of the Fondación Pro-Sierra Nevada de Santa María.

The map shows, in our view, the central traits of the mountain range: the topographic lines differentiate the massif topographically from the coastal plains and the Andes mountain range; the river basins that, through our relationship with the indigenous communities, we came to acknowledge as axes of spatiality in their territory, represent water, the most precious resource from the Sierra for surrounding areas. Finally, the map locates the "last primary forests" (Figure 9.1).

This map allows us to visualize the Sierra as "comparable to a mountainous island," as a microcosm characterized by biotic and cultural traits that make this massif unique. The notion that maps are created for specific purposes is not new. Borges tells in his *Universal History of Infamy* (1981), the tale of the prodigious map of an empire, so precise and complete that its dimensions coincided point by point with those of the empire. It was finally abandoned, however, to the rigors of the sun and the rains, showing that the usefulness of a map does not lie in the precision of what it represents, but in the representation itself. The power of cartography lies in that it produces representations of the world whose supposed neutrality masks the social order it represents, while at the same time legitimizing it. Maps, like any other historically constructed image, present, as Mitchell points out, "a deceptive appearance of naturalness and transparence concealing an opaque, distorting, arbitrary mechanism of representation, a process of ideological mystification" (1986: 8), what Harley (1992) calls the map's "political unconscious."

The Fundación Pro Sierra Nevada map took on an idiographic character that emphasized its biodiversity and a historical timeline that joined the present indigenous peoples with the ancient Tairona. This map quickly became an icon, and one could almost call it the Foundation's logo, appearing not only in documents but also in t-shirts and promotional brochures. The Foundation achieved great political success, gaining support from the President himself and quickly obtained important donations. Over the following years, the massif as a discrete unit delimited by topography became endowed with meaning as an emblem of the Colombia's tropical abundance, as the historic landscape of ancestral cultures, and as a place of social conflict and interaction.

Landscape and place

The vision of the massif and its diversity of landscapes that inspired the Foundation's rhetoric and much of the mystique in its members has a history that cannot be untangled from the invention of America as a pristine natural territory (Gerbi 1973; Denevan 1992). For Colombians, landscape has historically been one of the main attributes identified with national pride. Our "prodigious geography" and the "voluptuousness of its tropical nature"[8] have been valued in the poetic terms of the European pictorial landscape tradition, which responds to interests that are both aesthetic and instrumental (Thomas 1996; Berque 1994; Roger 1997). This tradition, just as it highlights the enormous wealth locked in our "vast and exuberant territory", proposes that it must be civilized in order to attain progress. These scenic images of tropical geography are anchored in a series of colonial notions about American history and culture.

The first of these ideas springs directly from the cartographic representation where on the horizontal surface of the map, the cordillera is transformed into a stratified succession of planes. Thus, the Andean topography appears horizontally segmented and classified as altitudinal strata. This vision hides the highly vertical understanding constructed by aboriginal societies about their territory. Ethnography and archaeology have documented the pattern of occupation known as "vertical control" or "vertical archipelago model" (Murra 1975; Langebaeck 1985). Leaving aside the fact that the notions of "control" and "model" are not entirely accurate to describe the relationships these groups have historically established with their surroundings, one can generalize by saying that their settlement was based on simultaneous management of several altitudinal levels, which permits access to an enormous variety of ecological niches. This vertical organization of territory is expressed in the fact that the system of river basins is an important social as well as spatial referent for these groups whose territories stretched, in pre-colonial times, from snowcapped peaks to the sea and the surrounding plains (Mason 1936; Reichel 1947; Cadavid and Herrera 1985; Serje 1987b)

For indigenous societies of the Sierra Nevada – and in general for the Chibcha cultural continuum in the Colombian Andes – the flow of water is the mirror of social and historical life of the populations: it is the axis of historical-territorial memory (Loochkartt and Avila 2004) as well as of people's mobility. Besides representing memory and thought, the system of lakes and basins constitutes the axis of an *embodied space*[9] that flows with the rhythm of everyday life and interweaves landmarks, connecting the mountains and ridges with the bays and peninsulas of the coastline, recreating the Sierra as a living body.

> The lake is the Mother's daughter, who put her on the mountains to give birth to rivers and streams. She was placed there by the Mother to have contact with the sea; she was put there to communicate through the river. The sea gathers all that the river carries. And from the sea rise the vapor and the froth, the clouds that return to the snow and the lakes, where it rains. Thus, there is constant communication between them.
>
> (Words of Mama Inkímaku from Makotama, OGT, 1997)[10]

The second colonial notion implicit in this representation of landscape is that of forests and jungles as "nature." In the European tradition, the cultural landscape can only be geometric, measured, exploited, and supervised (Deleuze and Guattari 1987; Hartog 1991). Landscapes that do not contain these "improvements" are seen as "vast and deserted wasteland." Jungles appear either as distant paradises, rich in exotic species, or as plague and fever-infested prisons, thus representing the archetype of virgin nature, both prior and opposed to civilization. They are valued as a source of promising species and knowledge, and as sites where the exquisite terror of extreme risks may be experienced (Harrison 1992; Dalla Bernardina 1996; Slater 2003). These images have guided all sorts of projects, from voyages of discovery and exploration to the most brutal extractive enterprises, and, certainly, conservation projects such as those of the Foundation.

The cultural appraisal that defines these landscapes as "wild" hides the role played by those societies that have historically inhabited them. Ethnology and archaeology have illustrated broadly the forest management practiced by its aboriginal inhabitants and have shown how, in great measure, its diversity is a result of their intervention. Jungle peoples have developed a type of temporal management of the territory in which various types of wild gardens or cultivated forests succeed each other (Rival 1998). These reflect the way in which groups classify and intervene in an area which, to outsider's eyes, appears generically as a "jungle". The *chagra* or plot, is not for these groups a domesticated area opposed to the wild forest (Descola 1986; Van der Hammen 1992). On the contrary, both are part of a continuum of plant successions in which the mature forest is one stage of this agricultural technique that is planned and managed on a timescale of over a century (Van der Hammen and Rodríguez 2000).

Efforts to study landscape as a cross-cultural reality usually naturalize the epistemological relationship that made the invention of landscape possible in the West, that is, the relationship between a contemplative subject and an object (nature) in which the observer abstracts a portion of the earth's surface, ripping it from its historical-geographical context in order to place it within a new one, that of science or aesthetics. While anthropologists have engaged in evoking the native's point of view, the status of landscape as an autonomous object has been assumed,[11] thus projecting the premises of the Western duality of nature-culture to the relationship that other societies establish with their surroundings (Descola 2005).

Perhaps in order to approach other forms of experiencing the environment, the concept of *place* may be more useful. According to Cresswell, "place is not just a thing in the World, but a way of understanding the world" (2004: 11). Anthropology has endeavored to understand the processes of both place-making and of dislocation. Thus, "anthropologists have come to worry less about place in broad philosophical or humanistic terms than about places as sites of power struggles or about displacement as histories of annexation absorption and resistance thus ethnography's stories about place are increasingly about contestation" (Feld and Basso 1996: 4). In the first sense, there has been an attempt to produce ethnographies of the perception and experience of specific places, examining the social relationships that produce them and seeking to describe and interpret the way in which people endow them with meaning. Discursive and performative practices through which places are formulated and experienced have been studied, since as Feld and Basso point out, "the intimate relationship between embodiment and emplacement brings the problem of place into close resonance with the anthropological problem of knowing 'local knowledge' " (1996: 9).

In the second sense, places are studied as processes immersed in national and global power relations. Attempts have been aimed at documenting and understanding processes of disarticulation and deterritorialization that accompany the creation of enclaves of inherent and consumer developments that make up the global economy, producing displacements, diasporas and exiles. If, as Lippard points out, place "is temporal and spatial, personal and political. A layered location replete with human histories and memories, . . . it is about connections, what surrounds it, what formed it, what happened there, what will happen there" (cited by Cresswell 2004: 40), the study of places and localities must also be the study of the struggles and the resistance for maintaining and constructing meanings for place.

The "black line": the sacred territory

The idea of limiting the Sierra Nevada as a territory defined by its topography – i.e., by a topographic line – ends up overshadowing the indigenous notion based on the verticality and the continuity of hydrographic basins in the lowlands. This topographic barrier, which separates highlands and lowlands, also has a long history in American geography and ethnology. This distinction has been assumed to be a "natural" geographic separation, opposing the *cultural* landscape of the cordillera, whose temperate climate is considered the natural habitat of peoples of industrious character – and the *natural* landscape of the warm lowlands inhabited by backwards and lustful peoples incapable of taming the jungles and savannas. This opposition constitutes the geographical correlate of the linear history of human progress, which begins at the most primitive, represented by the savage tropics, and arrives at the most civilized, represented by a temperate, urban Europe. This geographical characterization has been especially important in the history of Colombia, since the social hierarchy of its inhabitants and regions was established based on the distinction between the lowlands, where people are subject to the nefarious influence of the torrid climate, and the "elevated," cultured and industrious groups of the Andean region. In the case of the Sierra Nevada, this distinction operates in powerful but subtle ways: it sets in opposition the coastal groups and their corrupt and ignorant administrations and the "elder brothers," heirs of one of the great American civilizations and elevated spiritual traditions.[12]

The delimitation of the Massif based on its topography also reflects the vision of territory from Kogui cosmology, formulated in the ethnography of the Sierra:

> The Universal Mother, essence of all that is created, dug her spindle into the snowcapped peaks, from her spindle she pulled the thread and drew around it a circle; the region it contained she gave to her eldest children, so they and their descendants may inhabit there. The peaks where Mother dug her spindle, became the heart of the world . . . and the circumference traced by the Mother has been called the *black line*, and it has numerous guardians who appear to be of stone.
>
> (Botero* 1986: 4)[13]

Numerous cartographic attempts have been made to fix the black line, this mythical circumference, exerting pressure on indigenous organizations to identify the points that constitute it (Gil 1994).

Attempting to translate mythical space into topographic space implies ignoring a crucial dimension of indigenous knowledge of the Sierra Nevada. On the one hand, concepts and categories that have different meanings and connotations depending on their context are used. Even everyday discourse is woven through a network of analogies and semantic references that imply several dimensions of reading and interpretation simultaneously. The elders intervene by identifying the connections between everyday life and the contents of the narratives that safeguard historic memory (Gil 1994, 2007). This stance implies that the meanings of stories and myths are related to contemporary problems and political situations. To understand myth as an ahistorical reference, fixed by the ethnographic text, is part of a representation of indigenous peoples as natural as both natural peoples and as naturalists ("naturales o gente en estado de naturaleza por un lado y naturalistas, o conocedores de la naturaleza, por otro Descola hace referencia a ambas representaciones") (Descola 1985) who have preserved a sacred relationship with the cosmos thanks to isolation, or who resist Western intervention. This representation argues these societies – equated with the infancy of humanity – "live in a Cosmos regarded as sacred and participate in the sacred cosmic reality manifest both in the animal and in the plant worlds", as put by Mircea Eliade, who places them in explicit opposition to "modern societies who live in a de-sacralized Cosmos" (1967: 22).

The vindication of the spiritual-religious nature of indigenous thought and way of life – or in general of local and aboriginal groups in many parts of the world – has become in recent years a strategic axis for identity construction. The sacred has been incorporated into the repertoire of emblematic traits with which these identities are staged. In fact, various institutions, including the World Bank, began to incorporate the dimension of "the spiritual" into their policies for indigenous territories.[14] This process implies the objectification of the spiritual and sacred through a process of "scientisation," similar to that to which traditional knowledge must be subjected in order to be incorporated within the framework of technical and objective practices for planning and development (Agrawal 2002). This process entails the reification and reduction of the sacred, invoking essentialist categories. The reduction of indigenous historical-territorial memory to a "black line" drawn on maps implies tying the history of indigenous societies to a locality and to an image (Appadurai 1988; Malkki 1992; Rodman 2003). Here, the starting point isn't just a conception of the Massif as a "taken-for-granted setting to situate ethnographic descriptions" (Low and Lawrence-Zuñiga 2003: 15), but also a forgetting that places are "politicized, culturally relative, historically specific, local and multiple constructions" (Rodman 2003: 205). The essentialist appropriation of the "black line" blindfolds itself to the dynamic and conjectural nature of place. By fixing the sacred places in Cartesian space, the *interanimation* process through which region is built (Basso 1996: 55) is thus objectified.

The cartographic notion of the "black line" stems from a monological and essentialist vision and not from a confluence of voices and landmarks that create connections with the surroundings (Kahn 1996), like the one taking place today in the Sierra. The area of the Historical Park has become the stage for an indigenous resettling process. Ramón Gil, spiritual adviser of the indigenous organization who led a migration to this area from distant basins where their "traditional" territory is supposed to be located, tells that he decided to come here with his people because "my father's grandmother had told me that I had to come to the Nakulindue and Doanama basins to recover, to look for the wisdom, intelligence, knowledge, culture; because in Doanama two books are set in stone . . . That is why I came; in search of that book (Gil 2007).[15] The process of place-building in this new locality is related for this group to the recreation of knowledge and memory. Reviving "ancient" names creates a new toponymy: the Guachaca and Buritaca basins are renamed *Nakulindue* and *Doanama*. By thus naming them, this locality is connected to history and cosmology. The stone-books to which

Ramón Gil refers are the vehicle to recreate the past: the "books set in stone" allow this community also to "inscribe" their geography, while at the same time rebuilding the territory as living space. The stones link the community with the knowledge of the elders and with the world of the ancients. For the present, they are landmarks that ratify their right to occupy a locality even if it is not part of the "traditional territory" fixed by ethnography. In addition, they are read as referents of the knowledge that reiterates the identity of their people and their territory.

The massif: region of conflict

The indigenous territory appeared to the Fundación Pro Sierra Nevada as a world threatened by chaos and violence, which called for public intervention. To this end, a proposal was directed at local and regional administrations to conduct a *Diagnostic of the Sierra Nevada de Santa Marta*. Even though

> the work had been initially conceived as an inventory of natural resources leading to a management plan, the seriousness of the social situation, which turned increasingly violent, made evident the need to give priority to the socioeconomic situation.
>
> (Fundación Pro Sierra Nevada 1988a: 2)

> el trabajo se había concebido inicialmente como un inventario de recursos naturales conducente a un plan de manejo, la gravedad de la situación social, la cual se hizo cada vez más violenta, hizo clara la necesidad de dar prioridad al conocimiento de la situación socioeconómica.

The "diagnostic" made its main objective

> to delve into the causes of the situation at the Sierra, characterized . . . by an intensification of social conflicts, the absence of updated information in several areas, the lack of coordination in planning and execution by state agencies.
>
> (Fundación Pro Sierra Nevada 1988a: 2)[16]

This work included an "institutional diagnosis" that sought to determine the need for government services, and a "social diagnosis," which sought to show the "current situation of the population" and "recent socioeconomic processes" through a historical study based on archives and secondary sources, an oral history of colonization, and a study of the conditions of the borders of the indigenous territories.

Feld and Basso remind us that "geographical regions are not so much physically distinct entities as discursively constructed settings that signal particular social modalities" (1996: 5). In this sense, the Sierra Nevada is constructed as a region through this "diagnosis". It shares the central features of a set of "frontiers" that are perceived as fronts of resistance to the expansion of progress and civilization. Analysts refer to them as "empty spaces", "war territories," or as "internal borders". They have historically been imagined through the same set of representations. On the one hand they are seen as a promise of enormous wealth and opportunities. On the other, they represent danger and risk. They are seen both as strategic for development – for their biodiversity, water and mineral reserves – and as a threat to national stability – because of their seemingly constitutive violence, rebellion, and drug trafficking.[17]

The *Diagnostic* represents the Sierra through basic traits: it emphasizes the absence of the state and its irresponsible management by coastal administrations, by virtue of which the Sierra is subject to the "law of the jungle" and experiences "extreme poverty that survives in a climate of violence and injustice".[18] It is presented as a territory whose

> image of geographical anomaly, shrouded in clouds and isolated by steep slopes has made it throughout history a setting for illusions and utopias, and a refuge for those who have little to lose through a new adventure,
> (Fundación Pro Sierra Nevada 1988a: 16)
>
> imagen de anomalía geográfica, envuelta en nubes y aislada por fuertes pendientes la ha colocado a través de la historia como escenario de ilusiones y utopías, y refugio de quienes poco quieren perder con una aventura más,

and where illegality reigns, since

> the tradition of contraband created a culture, forms of organization and infrastructure that were profitably adapted to the production and commercialization of marijuana. Contraband is a constituent part of the prevailing social structure in the region.
>
> la tradición del contrabando creo una cultura, unas formas de organización y una infraestructura que se adaptó rentablemente a la producción y comercialización de la yerba [sic]. El contrabando es parte constituyente de la estructura social imperante en la región.

as a place where "there is a culture where the illicit is legitimate, because State action seems to stop at the piedmont" (se desarrolla una cultura donde lo ilícito es legítimo, porque la acción del Estado parece detenerse en el piedemonte) (Fundación Pro Sierra Nevada 1988b: 51).

The *campesino*-settlers are represented as predators: "The settlers systematically destroy their environment; and in their wake lay waste to all regional forest preserves and fauna since they lack the knowledge that would allow them to properly manage resources, [this] has meant that a process of erosion in the area is irreversible" (Mayr 1984: 7).[19] They are represented as a group determined by a historical fatalism:

> violence in the Andean country produced a settler "without God, King, or Law", familiarized with violence and willing to use it to achieve their

> goals; settlers organizations were born out of violence and they perpetuated it in the Sierra as a means for social action, unleashing in those affected a reaction identical in its nature.[20]
>
> (Fundación Pro Sierra Nevada 1988b: 51)

Emphasis is placed in portraying this group of settlers as "prone to violence . . . that can't and does not know how to avoid it. Besides, they have experience in organizing it and exerting it socially. The guerrilla has thus an open door" (Fundación Pro Sierra Nevada 1988b: 52).[21]

The indigenous society is also presented in the *Diagnostic* through a history of refuge: "the Sierra was a marginal area covered by extensive wooded zones that separated the incipient and slow development of the lowlands from the life of the surviving aborigines in the highlands."[22] It points out that "out of these four groups, three subsist precariously, since the Kankuamo are culturally extinct and . . . the Wiwa are in serious risk of cultural extinction" (Fundación Pro Sierra Nevada 1988c: 1–2),[23] and these societies are presented through a gradation from the most traditional to the most "acculturated:"

> The Kogui are the most zealous guardians of their tradition and culture, they maintain great respect for their authorities, very few of their women speak Spanish and it is the group in whose territory money circulates the least. The Arhuacos . . . are not a homogeneous group: on the one hand there are those traditional sectors with similar characteristics to the Kogui; there is a second sector with greater understanding of the country's sociopolitical reality, interested in developing a market economy and . . . the third sector no longer wears traditional clothing, speaks Spanish and has clearly peasant customs and interests.
>
> (Fundación Pro Sierra Nevada 1988c: 3)[24]

This analysis, which does not concern itself with historicizing the representations of these societies,[25] ends up taking legitimacy away from the "mestizo" groups:

> since they are settled inside the reservation, they modify in some cases their position depending on their interests. They present themselves as indigenous before the government and before the indigenous communities they present themselves as white . . . they have co-opted the representation of indigenous communities before State institutions in order to favor their immediate interests, channeling resources that were destined to the indigenous communities for their own benefit.
>
> (Fundación Pro Sierra Nevada 1988: 3)[26]

The diagnostic document fulfilled the function of defining the traits of the Sierra Nevada as a planning region. Scott has shown that implanting any form of order "requires a narrowing of vision . . . [that] brings into sharp focus certain limited aspects of an otherwise far more complex and unwieldy reality" (1998: 11). The exercise of diagnosing – a hygenic term used to designate the reading of a situation aimed at formulating policies, programs, or legal arguments – allowed the Fundación Pro Sierra Nevada to identify and characterize the aims of government action; they portray the Sierra as a marginal and wild space governed by the will of the strongest and, above all, it shows its inhabitants as groups defined by scarcity and anarchy.

De Certeau (1984) pointed out that the power to territorialize is materialized through classification, delimitation, and separation, which may be considered as spatial strategies. One condition that makes these strategies possible is the practice of contextualization. The need to "put things in context" is a necessary *sine qua non* condition of social analyses. One of the main objectives of the social sciences is precisely that of situating the data, facts, and phenomena that they study "in context": we begin with the premise that a phenomenon or process in focus can be described and interpreted appropriately only if one looks beyond the event itself and if other phenomena, which provide resources to do this, are put in focus as well. In this regard, context, as an object of study, is the result of a multiple process of selection and interpretation. What is defined as a problem, together with what is deemed relevant as its explanation, depends on the way in which context is evoked and produced (Duranti and Goodwin 1992). Context takes shape through connections established between them and, naturally, through their disconnections (Dilley 1999). It is perhaps here where the etymology of the word context becomes particularly important, as it is derived from a textile metaphor in the Latin verb *contextere:* to tie, to weave, to join the threads of a fabric.

Context is usually understood and expressed as a spatial reality through images such as "scene," "backstage," and "setting," which relate to the idea of a theatrical spectacle as a metaphor for social organization. Geography and representation merge in the symbolic and ideogrammatic tradition of the *forma orbis* in which a single image represents the world, showing its conditions in an abstract and systematic manner (Alpers 1987). Thus, the historical-territorial reality is transformed into a scenic space on which to act. The idea of context as description of the world – of a world that gives meaning to an event or focal process – also proposes an image that appears simultaneously as a model, an ideogrammatic description, and a setting for action. It contributes to the illusion of the transparency of space (Lefebvre 1991: 28), showing it through a relatively objective description of a given reality, at the same time as it legitimizes and naturalizes social-spatial relationships.

The production of context as a spatial practice is constitutive as well as a result of social action. By defining the context in which a relationship or process takes place, the nature of the setting is determined, as are the antecedents and relevant actors and their roles. Thus, contextualizing establishes the enabling conditions for legitimizing certain forms of intervention. The assumptions underlying the creation of "frontiers" such as the Sierra Nevada – characterized by wild landscapes, the absence of state, "the law of the jungle," uprooted populations prone to illegality, neutralized natives – turn them into an "inversion" of formal spaces where

> the "others" of the master subject are marginalized and ignored in its gaze at space, but are also given their own places: the slum, the ghetto, the harem, the colony, the inner city, the third world, the private. These places

> haunt the imagination of the master subject, and are both desired and feared for their difference.
>
> (Blunt and Rose 1994: 16)

At the same time, they legitimize and make possible the forms of action and administrative figures through which these spaces are managed and intervened. Disembedding landscapes and social groups from their historical-geographical continuity in order to see them as "frontiers" gives rise to negative representations: militarization and paramilitarization, slavery, debt-peonage, trafficking, prostitution, intensive and extensive exploitation of licit and illicit resources. Here, anything goes.

Epilogue: zone of intervention

In large measure, lobbying by the Fundación Pro Sierra Nevada made the Massif into an area of state intervention. During the 1990s it became a "special district" of the National Rehabilitation Plan, a governmental initiative directed at regions suffering from "conditions of social conflict characterized by their marginality and by the absence of the State" (SIP 1990).[27] Since 2000, it become one of the nine "priority areas of intervention" for the "reconquest of the territory," an essentially military strategy against drug trafficking and guerrillas. Its objective is to consolidate "the opening of these zones to development and to the global economy" (CCAI 2006).[28] The aim was to eliminate unruliness and allow profit from their resources. The area has now also become a theater for military operations. A "high mountain battalion" was stationed in the heart of indigenous territory.

After the state adopted the Sierra as an intervention zone, the Fundación Pro Sierra Nevada focused on biodiversity conservation. The foundations had already been laid: the Massif is considered to be a microcosm of the various tropical mountain ecosystems; it provides a habitat for numerous endemic species and is the home of cultures represented as homeostatic systems adapted to their environment (Uribe 1988, 2006; Ulloa 2004). The internationalization of the Fundación Pro Sierra Nevada and its positioning in the world of global environmentalism (as an active member of the UICN, the Global Biodiversity Forum, etc.) spearheaded a series of interventions that have turned the Sierra into a setting for international environmental action. Paradoxically, this type of involvement was aimed at those sectors of the population that have the most limited impact on the ecosystems, i.e., local populations. Those groups whose decisions and projects have most decisive effects on land ownership, landscapes, and the living conditions of its inhabitants are beyond their reach: investors, agents of agro-industrial projects, power blocs linked to coca and the paramilitaries, and even the technocratic sector of the state, all of which are adamant about "competitiveness" in global markets. While conservation was reduced to a problem of protected areas and local communities, the Sierra was opened to global investment.

Gradually, a "land use" model emerged that delimited conservation areas, generally assimilated into the indigenous area (which overlaps a Natural Park), and the rest of the Massif as an area of development. Both are geared to give way to a variety of economic initiatives. The former are reserved for the consumption of nature through ecotourism, adventure tourism, and recently "spiritual" tourism, which sells ritual ceremonies of Indians from the Sierra. Areas that are not protected are geared to the "hard" lines of regional development such as agro-industrial production, the intensive exploitation of resources, and the construction of large infrastructure projects.

Thus, a distinction is established in the Sierra between a "sacred" space, aimed at conservation of biological diversity and its "ecological natives" (Ulloa 2004) and a space prepared for the development of the global economy. They are not, however, in mutual opposition: they represent two faces of the same coin. They constitute a *geography of management*, whose enabling condition is the *geography of the imagination* that stems from the practice of contextualization woven with the routes traveled by our work group. Some of these roads allowed us a glimpse into a world in the Sierra. Others opened the way for the establishment of new connections between the Sierra and the world. Somewhere in the crossroads lies a utopia.

Notes

1 This decade is known in Colombian history for the bloody clashes between the Liberal and Conservative parties, triggered by a popular uprising known as *El Bogotazo* and lasting until the political agreement between the two traditional parties to alternate power.

2 Several works have accounted for the way in which space and spatiality – as objects of study – have been problematized and conceptualized in anthropology, summarizing theoretical debates: Ellen 1993; Hirsch and O'Hanlon 1995; Low and Lawrence-Zuñiga 2003.

3 Here I refer broadly to the concept of 'practical mastery' proposed by Bourdieu (1977).

4 What Low calls social production of space, "the processes responsible for the material creation of space as they combine social, economic, ideological, and technological factors" and social construction of space, "the experience of space through which peoples social exchanges, memories, images and daily use of the material setting" (2000: 128).

5 Works produced by this group are marked by an asterisk (*).

6 Trouillot distinguishes between "*location* as a place that has been situated, localized if not always located. . . . *locale* as a venue, a place defined primarily by what happens there: a temple as the locale for a ritual . . . [*Locality*] as a site defined by its human content most likely a discrete population" (2003: 122–3).

7 The original reads: "Primero estaba el mar. Todo estaba oscuro. No había sol ni luna ni gente, ni animales, ni plantas. Solo el mar estaba en todas partes. El mar era la madre. . . . La madre no era gente ni nada, ni cosa alguna. Ella era espíritu de lo que iba a venir y ella era pensamiento y memoria" (Reichel 1985: 17).

8 I place these phrases in quotation marks without attributing them to anybody, since they are part of common knowledge since the nineteenth century.

9 In the sense discussed by Low and Lawrence-Zuñiga (2003).

10 "La laguna es hija de la madre, quien la puso en los cerros porque desde allí nacen los ríos y las quebradas. Esta laguna fue puesta por la madre para tener contacto con

el mar, fue puesta para comunicarse por medio del río. El mar recoge todo lo que el río le lleva. Y desde el mar se levantan el vapor y la espuma, las nubes que van otra vez hacia la nevada y las lagunas, donde llueve. Así, hay una comunicación continua entre ellos."

11 Berque (1994) drafts a set of criteria to identify what can be considered *landscape cultures*: the existence of a word to designate it, a pictorial and a literary or oral tradition to praise it, and the existence of gardens of enjoyment. According to him, the notion of landscape is not universal, and in fact it only exists – with different meanings – within the Sino-Japanese and Western traditions.

12 In the republican history of Colombia, the cultured and refined elites from the Andean highlands are often opposed to the corrupt and backwards elites of the lowlands, especially the coast (Serje 2005).

13 The original reads: "La Madre Universal, esencia de todo lo creado, clavó su gran huso de hilar en los picos nevados, de él desprendió la punta del hilo y trazó a su alrededor un círculo; la región comprendida la entregó a sus hijos mayores para que allí habitaran ellos y su descendencia. Los picos nevados, el punto donde la Madre clavó el huso, quedó como corazón del mundo . . . a la circunferencia trazada por la Madre se le ha llamado la *línea negra* y tiene numerosos guardianes con apariencia de piedra."

14 See, for example, World Bank 2004.

15 In the original: "la abuela de mi papá me había dicho que tenía que venir a la cuenca Nakulindue y Doanama a recuperar, a buscar la sabiduría, inteligencia, conocimiento, cultura, porque en Doanama esta plasmado en una piedra dos libros . . . Por eso me vine en busca de ese libro".

16 "profundizar en las causas de la situación de la Sierra, la cual se caracterizaba [. . .] por la agudización de los conflictos sociales, la ausencia de información actualizada en diferentes áreas, la falta de coordinación en la planeación y ejecución de los programas de las diferentes entidades".

17 This set of regions, which constitute over half of the national territory, was the topic of a work (Serje 2005), where I discussed the way in which regional studies in Colombia contextualize the relationship between the state and these territories.

18 "una extrema pobreza que sobrevive en un ámbito de violencia e injusticia".

19 "la destrucción sistemática del entorno por parte del colono, que a su paso arrasa con todas las reservas forestales y faunísticas de la región por carecer de los conocimientos que le permitan un manejo adecuado de los recursos, ha generado en esta zona un proceso de erosión irreversible".

20 "la violencia en el interior del país creo un colono 'sin Dios, ni Rey, ni Ley', familiarizado con la violencia y dispuesto a usarla para alcanzar sus metas, las organizaciones de colonos nacieron de la violencia y la prolongaron en la sierra como medio de acción social, desencadenando en los afectados una reacción de idéntica naturaleza".

21 "es proclive a la violencia . . . no puede ni sabe evadirla. Tiene – además – experiencia en organizarla y ejercerla socialmente. La guerrilla tiene pues una puerta abierta".

22 "la sierra fue un área marginal cubierta por extensas zonas de bosque que separó el incipiente y lento desarrollo de las partes bajas de la vida de los indígenas sobrevivientes en las partes altas."

23 "de estos cuatro grupos hoy tres subsisten precariamente, pues los kankuamo se extinguieron culturalmente y . . . los wiwa se encentran en serio peligro de extinción cultural."

24 "Los kogui son los guardianes más celosos de su tradición y su cultura, conservan un gran respeto por sus autoridades, muy pocas de sus mujeres hablan castellano y es el grupo en cuyo territorio el dinero tiene menor circulación. Los arhuacos . . . no son un grupo homogéneo: por un lado se encuentran aquellos sectores tradicionales con características similares a las de los kogui; por otro lado un sector con mayor manejo de la realidad sociopolítica del país, interesados en el desarrollo de una economía de mercado y . . . el tercer sector ya no viste la manta tradicional, habla el castellano, y tiene costumbres e intereses netamente campesinos."

25 Uribe (1988, 2006), Langebaeck (2005), and Ulloa (2004) have proposed a critique of this ethnographic representation.

26 It reads in the original: "estando asentados dentro del resguardo modifican en algunos casos su posición al vaivén de sus intereses. Frente al gobierno se presentan como indígenas y frente a los indígenas como blancos, . . . han tomado la representación de los indígenas frente a las instituciones del estado para favorecer sus intereses inmediatos, canalizando recursos que iban destinados a las comunidades indígenas para su propio beneficio".

27 "condiciones de conflicto social, caracterizadas por su marginalidad y por la ausencia del Estado."

28 "la apertura de estas zonas al desarrollo y la economía mundial."

10 Spatiality and religion

John Corrigan

The investigation of spatiality and religion has a long history, from its roots in the ancient drafting of religious cosmologies, through early modern challenges to theologically inflected geographies, to recent cross-disciplinary experiments in theorizing religious space and place (Büttner 1973, 1980; Kong 1990, 2001, 2004; Park 1994; Knott 2005; Sopher 1967). Thinking about spatiality and religion, moreover, has evidenced, almost from its beginning, some measure of reflexivity about how it defines the phenomena it inventories and interprets. So, the *Geography* of Strabo, itself brimming with wonders incubated in religious imagination, was at the same time critical of the ease with which other writers made geography the handmaiden of myth. This criticism fell especially on Strabo's Greek predecessor Megasthenes, whose accounts of India Strabo at times dismissed as "going beyond all bounds to the realm of myth" (Strabo 1932: 15. 1. 57).

Jewish, Hindu, Buddhist, Muslim, Christian, and other cosmologies grounded space in the scriptures of those traditions. Writers representing those traditions sought correspondence between, on the one hand, the articulation and delineation of space in myth and, on the other, the environments – natural, social, emotional – in which people lived their daily lives. Trusting that spatial order was given in religion, they interpreted their everyday experience in such a way as to ensure its synonomy with that order. Medieval cosmographers – generally concerned in the West with demonstrating the reality of divine providence – sought to harmonize Aristotle's *Meteorologica* and other classical texts with biblical stories of creation. Vincentius of Beauvais's Christian *Speculum Naturale* (*c.* 1200), the tenth-century Muslim writings of Al-Muq-addasi, and the fledgling geographies of early medieval Celtic monastic academies, among many other efforts, advanced and complicated theological understandings of space. They accomplished this largely by mapping religious centers – places where holy persons had lived, miracles were performed, visions were realized, or religious truths revealed – and locating them in relation to heavenly territories and other places and events chronicled in scriptures or other holy writings.

In the sixteenth and seventeenth centuries, geography participated in the nascent scientific privileging of experience over revealed truth, and so moved steadily away from its previous role as a servant to religious dogma. Debates among various Christian groups about divine providence, the creation of the world, and teleology shaped geography in new ways, eventuating in Kant's definition of a geography separate from theology. But those debates also coalesced as new, specifically Christian geographies that appeared throughout the period, including those that emphasized the missionary responsibility of the Christian churches. Gottlieb Kasche's *Ideas about Religious Geography* (1795) employed the term "geography of religion" to identify the comprehensive spatial mapping of Christianity and its competitors (Livingstone, 1994; Büttner 1980; Park 1994: 10). Such geographies remained strongly theological in tone.

With the eighteenth-century development of systematic philological study of ancient Christian documents, the experimentation with more ambitious means of scriptural exegesis, and the growing fascination among Europeans with the archaeology of the eastern Mediterranean came the characteristic nineteenth-century emphasis on mapping biblical history. On the heels of Western colonial expansion, the "Holy Land" style of geography in the West was enlarged to enable mapping of the Asian subcontinent, North Africa, and, in turn, larger areas of those continental landmasses. Environmentalist explanations for religious belief subsequently became more prominent as geographers sought correspondences between local experience and items of religious faith. Landscape, climate, lifestyle (e.g., nomadic), and other factors were viewed in relation to religion, yielding the kinds of conclusions that say as much or more about the magisterial gaze of the European than about the people observed. The nineteenth-century French historian of religion and ideological gadfly Ernst Renan, at times strikingly reflexive in his writing and at others as transparently naïve as his sometime collaborators, had his doubts about some such correspondences. Displaying the kind of critical perspective that historically has emerged with some regularity to inform writing about religion and geography, he wondered about the claim that "Le désert est monothéiste," that is, the notion that monotheism was the logical response to the experience of smallness under the vast, starlit night sky of the desert. For Renan, "en vérité, le désert a véhiculé toutes sortes de religions: le chamanisme de Toungouses, le bouddhisme des Mongols aussi bien que le monothéisme musulman." By the end of the nineteenth century, in an intellectual climate increasingly shaped by evolution, natural law, materialism, and hardening canons of scientific inquiry, Western geographic survey of religion drifted from its role as argument for the superiority of Christianity (Sopher 1967; Livingstone 1994; Deffontaines 1948: 130).

The study of religion and space developed through several phases during the twentieth century. As a number of scholars have argued, environmental determinism – religion as the product of geographical factors (and especially climatological and topological factors) – carried over in various ways from the nineteenth century (Kong 2004; Levine 1986). But the emergence of the comparative study of religion in Europe and America and particularly the growing interest in defining religion began to redirect geographical analyses in important ways. Max Weber's emphasis on the manner in which religion shaped economic, social, and legal institutions was adopted as a guiding principle by many as they interpreted data relevant to religion and space. Two issues led the way among scholars researching religion and geography. The first was the question of what sorts of phenomena ought to be considered "religious." Pierre Deffontaines's *Geographie et religions*, published in 1948, Paul Fickeler's (1962) "Fundamental Questions in the Geography of Religion" and David Sopher's (1967) *Geography of Religions* together identified those elements that have been taken by most geographers since about mid-century as crucial to any understanding of religion. For Deffontaines, place of dwelling, demography, exploitation (agriculture, industrialization, animal life), movement (circulation of people, goods, the dead), and lifestyle (food, work, seasonal calendars) were crucial. Fickeler contributed the additional elements of ceremony and religious toleration. Sopher added human ecology, elaborated on the importance of pilgrimage, and drawing on historian of religion Mircea Eliade, advanced the central notion of sacred space. In none of these seminal works was religion viewed as simply either the product of environment or the motive for landscape change. Some measure of give and take between religion as prime mover and religion as socially shaped was redolent in these studies and that spirit of dialectics has proven durable up to the present. Manfred Büttner's emphasis on the religious body – the organized cluster of practitioners of a religion – as an active mediating middle ground between the environment and religion encapsulated this aspect of previous research (Büttner 1980).

Lily Kong, Chris Parks, Gregory Levine, Danièle Hervieu-Léger, and others have offered schemata for organizing the trends in research on religion and space since approximately the mid-twentieth century (Kong 2004; Parks 1994; Levine 1986; Hervieu-Léger 2002). Such overviews agree that the study of religion and space in recent years has become much more complex, and that such complexity is the product of more ambitiously interdisciplinary inquiry and the maturing of a pointed reflexivity on the part of researchers. What has been lacking in these otherwise thorough overviews – written by geographers – is familiarity with the kind of issues and problems that have driven debate within religious studies in the last few decades and in some cases eventuated in redefinition of religion. In other words, there is to a certain extent a specialized discourse, represented in the discussions among scholars working in religious studies, that foregrounds some themes overlooked by research steeped in the agendas set out in geography journals. Consequently, while noticing ways in which research among religionists overlaps with that of geographers, it is worthwhile to explore outward from geographers' expertise with spatial analysis to the central concerns of religion scholars regarding space and place.

Invisible worlds

Central to the ongoing project of describing and explaining religion is investigation of the invisible worlds imagined by believers. In their own ways, Dante Alighieri, the Mormon founder Joseph Smith, and the authors of *The Tibetan Book of the Dead*, among many other writers, were surveyors of landscapes made real to Christians, Mormons, Buddhists, and others through religious faith. There is a rich history of the geographic exploration of those invisible worlds, and for believers, religious history is as real as the history of the exploration of landforms, peoples, languages, flora, societies, climates, and every other aspect of life on Earth. Religion conflates the visible and invisible, the world of the senses and the world of the imagination. Accordingly, the space that scholars study when they study religion is a territory that often is nondefinitive, protean, multivalent, temporally ambiguous, irregular, and by definition ultimately unchartable. Folded together, as it always is, with physical space (the space of geography), religious space can take the form of Atlantis, Mount Olympus, a mosque, a cemetery, a dining table, a nation, a social class, hell, heaven, or a jazz room. The investigation of space on the part of religion scholars consequently tends to a style of inquiry that focuses at every step on how the symbolization of space is related to the occupation of space. That is, the meaning of space itself is rarely transparent, and it is only through a process of gauging a community's investment in the imagined and invisible territories of religion that understanding of religiously inflected dynamics of everyday life – gender, nationality, ritual, ethnicity, politics, sexuality and so forth – is possible. Imagined worlds are built with materials drawn from the experience of earthly environment: for Eskimos, hell is a place of frigid darkness, while for Jews it is a place of intense heat and for Christians (*à la* Dante, who was exiled by the Pope) it is a territory marked by the diabolical inventiveness of community leaders, the caprice and meanness of demons who hold authority in the ordering of infernal society. By the same token, earthly political institutions, for example, can embody – according to a number of religions – the dynamics of dominating power epitomized in the collective histories of heavenly and hellish denizens, and the sensual and emotional aspects of the earthly experience of God can follow – as in Puritan conversion narratives – models observed in visions of the reign of God over the departed faithful.

The study of space and religion, then, involves first of all a willingness to incorporate data drawn from the testimony of those who see, hear, taste, touch, and smell places that do not show up on the academic geographer's map of the world. Religious space is always polylocative, and among the places that might be associated one with the other are those that are made real in visions. The pious believe that persons travel back and forth between those places, and that spaces paradoxically can overlap or bleed one into the other, and be simultaneously inhabited. This religious mentalité rests upon a determination to challenge the authority of boundary, however that boundary might be warranted. One way of observing this notion is through examination of reports of human visits to supernatural worlds. The literary genre of the apocalypse, especially, is characterized by accounts of heavenly tours, and ascents and descents into the various supernatural realms. *The Book of Dream Visions,* a text known and cited by early Christians, and the canonical Revelation to John, for example, are replete with supernatural personages who, like the seer, travel from one place to another, involving themselves in the social and political life of both the supernatural and earthly provinces, and through their activity demonstrating the ease with which geographic distinctions can be subverted and even trivialized. As Leonard Thompson (1991: 117) observed in his attempts at mapping apocalyptic worlds, "boundaries are soft and permeable, open to passage. Therefore, distinctions between objects in the seer's world are not absolute and categorical; they are relative, with one object blending into the next." For the believer, the invisible world remains somewhat indeterminate and fluid, even in the midst of parochial "thick description" of its features. Its relationship to earthly geography, while embraced with absolute certainty by religionists, likewise is ambiguous. Finally, places on earth, for those invested in the religious ordering of the cosmos, are similarly conceived as both bounded (e.g., there is such a place as Mecca) and unbounded (i.e., Mecca is spatially present and encompassing to Muslims as they face it to pray) (Schimmel 1991).

Pilgrimage

Mecca is the primary site for pilgrimage for Muslims. It represents events in the history of the religion's founder, the historical development of the religious community, the aspirations of Muslims, and the linkage with the invisible world through its location opposite the heavenly Ka'aba. It is the point through which the world's axis passes. It is, in the words of historian of religion Mircea Eliade (1961), the location of the *axis mundi*, the spatial center from which the points of the compass extend and the temporal center which is both beginning and end. The sacrality of the Ka'aba at Mecca authorizes the ongoing theological project of Islam, as well as the social and political and jurisprudential aspects of Islamic life that flow from theological investigation. Mecca as a site of religious power is pre-eminent, stable, permanent, unchanging, and powerful. It is, at the same time, unstable, shifting, and ambiguous.

The meaning of Mecca, like the meaning of all religious sites, is constructed in religious practice. Because of its commanding power, it is a place where persons wish to be, a place that can be relied upon to spiritually inform and nourish in the most profound ways. But, as Thomas Tweed (2006: 123) recently has written, "religions are not only about being in place but also moving across." Religions are constructed and maintained through simultaneous emphases on the power of bounded place and the importance, the necessity, of transgressing boundaries. The meaning of Mecca emerges through the interplay of two seemingly conflicting processes: the ongoing verification of the site as a fixed center of power, and the devotional activity of Muslims whose experience of Mecca is as much the experience of movement across boundaries – national, natural, social, ethnic, gendered – as it is the emotionally and intellectually certain embrace of a bounded geographical point. In short, the transgressive act of pilgrimage is crucial to making meaning of Mecca. The experience of Mecca is the experience of arriving and staying long enough to see, smell, touch, hear, and taste it, and in so doing to submit to its authority. At the same time it is the experience of leaving one's home, one's nation, one's continent; crossing mountains and seas; stepping outside the world of one's ethnicity to collaborate with persons of other ethnicities; abandoning the landscape of class to mingle with others who have done the same; trade gender segregation for some measure of joint worship in the vicinity of Mecca; allow lines marking political difference to blur; surrender familiar bodily habits of eating, sleeping, and moving; and, for many, cross boundaries defining sectarian debate. At its most obvious, pilgrimage is travel from one place to another. Considered in its complexity, it is a multilayered ritual of confirming the sacredness of a place through a process of subverting, through physical travel to that place, the very idea of boundary that undergirds the authority of place. For a pilgrim, the experience of being in Mecca is intertwined with the experience of getting there and returning. Consequently, the experience of Mecca radiates across all of the landscapes that the traveler has traversed in order to arrive at the destination. Mecca consequently is present to the pilgrim not just in the act of facing east during prayer, but through immersion in the social, ethnic, gender, political, and national landscapes that contextualize and define everyday life. Theologically mapped as the flowing of landscapes one into the other, polylocativeness is a central feature of religious life (Mary 2002).

The devotional exercise of pilgrimage – whether to Mecca, Lourdes, Guadalupe, or Mount Kailash in Tibet – illustrates something of the manner in which spatial ambiguity and spatial definitiveness cooperate in the workings of the religious imagination. Not everyone goes on pilgrimage, however. Most religious persons do not visit sacred sites in Arabia, or Mexico, or Tibet, or other places, nor do they engage much in travel to less esteemed shrines located closer to the communities they inhabit. Many religious persons, however, partake of a belief system that situates them in relation to idealized pasts or expected futures. The obvious instancing of the latter includes heaven, hell, purgatory, or other landscapes of the afterlife similar to those already mentioned above. The former – the memory of what has gone before, the places, people, societies, material culture, and so forth – can be understood more precisely with reference to migration.

Religion deploys a sophisticated rhetoric in attracting adherents and in keeping them as members of the community of believers. A central component of that rhetoric is its appeal to the ancient origins of the religion. Religions have cosmogonies – stories about the creation of the world – that are crafted in ways to illustrate the conformity of religious practice with patterns of thought and action characteristic of the earliest living persons and with the divinities who oversaw their lives. Believers, through their participation in rituals and other devotional performances, remember that terrifically distant and receding past, including all of the various landscapes of the past, and seek to recreate them in the present. Religion is about the ongoing return to the past (because to forget the past is to lose faith). And the impulse to return is manifest in religion in many ways. While we might take emulation of the lives of ancient prophets and teachers as a crucial category for recovery of the past, we ought also to appreciate the manner in which the past is memorialized through more recent events, and especially how the fact of migration, which has been a central aspect of religion throughout its history, has proven instrumental in shaping religious sensibility and directing religious devotion.

Migration

The study of religion and space in the West has long noticed the scattering of Jews. This diaspora has contributed importantly to the Jewish mythologization of areas of the eastern Mediterranean as sacred ground. It has urged upon Jews a diligence in remembering the place from which they were dispersed and kindled imagination and desire to enrich the meanings of those places through theological reflection, imaginative writing, and the invention and refinement of religious performances. Jews, like other religious groups, have lived in diaspora, and that experience, while not one that a person might choose, has proven useful in creating an understanding of place thick with religious meaning.

Migration, whether it is forced, as in the case of the Jews, or as migration that takes place with some measure of assent on the part of the migrants, has been crucial to the historical development of religion. The mental gymnastic of folding spaces one into another that is characteristic of religious conceptualizations of reality has emerged from the experience of migration as much as it has developed out of thoughtfulness about the cosmic ordering of visible and invisible worlds, the locations of the *axes mundi*, and the ambiguity of borders. So important is it to the vitality of religious life, in fact, that religious communities have on occasion invested in memories of migrations from imaginary elsewheres. Chantal Saint-Blancat (2002: 138–9), remarking on that phenomenon, pointedly explained that such situations evidence that "mobility, far from being something to be suffered, is managed as a resource." Diaspora is a "place of tensions, of continuous re-adjustments, a space of fragmentations and of unifying processes, symbolically as well as on the level of social practices."

Religion renders deterritorialization as reterritorialization. In so doing, it establishes frameworks for maintaining identity and for defending theological claims for cosmic order and moral reckoning. As Danièle Hervieu-Léger has observed, religion exploits the creative opportunity for interrelating "detachment from concrete territorial inscription, brought about by migration, and mobilization of an idealized territoriality that provides the raw material for reconstructing identity" (2002: 104). There are a number of scholarly investigations of this process in modernity that stress the paradox of the migratory experience as represented in religious mythology and devotional practice (Saint-Blancat 2002; Dianteill 2002; Mary 2002; Levitt 2004). We should remember, however, that migration is not merely an artifact of modernity. It is – as in the case of the Jews – a long-standing fact of religious, ethnic, and national history. And religions have over many centuries drawn upon sacred geographies to establish new understandings of space as their adherents migrated from one territory to another.

Reterritorialization, when managed primarily through appeal to a religious worldview, frequently can involve the construction of plural religious identities. Robert Orsi, in analyzing the emergence of an Italian-American Catholic community in Harlem in the nineteenth and early twentieth centuries, detailed the ways in which an assortment of identities, interwoven under the umbrella of religion, took shape among immigrants through the utilization of religious practice. Those identities included gender, nation, family, neighborhood, class, language, and denomination, among others. Narrating the immigration of Italian Catholics to New York, and interpreting their efforts to come to terms with life in a new place, Orsi noted how the transitional process was advanced through immigrants' adaptation of an Italian religious festival celebrating the Virgin Mary.

Characterized by bold displays of piety, the annual *festa* in Italian Harlem featured the theme of healing. During the period of the *festa*, members of the community would seek healing from the Virgin, and, in bold displays of piety, signal their devotion to God and Mary as well as to the community, family, gender roles, the old country, and neighborhood, among other things. They also registered their resistance to what they believed was antithetical to their well-being, and perhaps most importantly in this regard, to the authority of the Catholic hierarchy in America, who were embarrassed by the persistence of a decidedly local Italian cast to the Catholicism of the immigrants in Harlem. Those displays included acts of humiliation such as travelling a church aisle on one's knees or with one's tongue on the floor, as well as the ritual deployment in church of wax body parts resembling those for which the supplicant sought healing. The sacred space of the church served as a microcosm for the immigrant society as whole: the healing of wounds (physical and emotional), the remembrance of the past, the reaffirmation of the places of women and men in society, the centrality of the family, and so forth – all of these identities were revivified, adapted, and resolved each year at the *festa*. Crucial to understanding the nature of the festival is the fact that the entire undertaking was predicated on participants' awareness of their inbetweenness in the American setting. Catholic but at odds with American Catholic leaders, Italian but living in America, committed to traditional family structures but forced to make peace with different family models, seeking healing but cultivating their woundedness – those persons imagined themselves as in two places at once, and, aided by religion, traveled conceptually between those two places and the identities associated with them (Orsi 1985).

Similar investigations of the religion of immigrants to the United States have detailed the capability of religion to represent space in complex ways, and, especially, to massage it conceptually in the service of adapting identity. Kristy Nabhan-Warren's study of Marian apparitions among Mexican Americans and Luis León's tracking of the development of bilocality in devotions to the Virgin of Guadalupe (i.e., Mexico City and Los Angeles) represent the current direction on scholarship concerned with religion and migration in their focus on the religious cultivation of the haunted life. That haunting takes place largely as the outcome of the simultaneous affirmation of the location of sacred space and the dissolution of the boundaries that separate it from other places. In the case of Mexican American Catholics in Los Angeles, that process was represented in 1999 by the display in Los Angeles of the cape of Juan Diego (to whom the Virgin appeared in 1531), upon which the Virgin's image had been miraculously emblazoned (Nabhan-Warren 2005; León 2004).

Ritual

Sacred space is kept sacred through ritual. Ritual delineates sacred space and at the same time provides for its portability and malleability. Ritual performances, as Victor Turner (1969) explained, can exploit the valences of space in such a way as to reverse the social landscapes of class, status, gender, office, and other structural elements of social order. Ritual likewise can transmogrify spaces both large and small through its capability to transplant objects, bodies, light, sound, and community from one location to another. Whether we are speaking of the wine inside a chalice becoming blood in Christian ritual, or a shaman returning from the invisible world with a soul that had drifted from its corporeal body, or the real linkage of a holy Muslim personage with his drawn image, ritual denatures, mutates, and restyles space to permit such feats. Through prayers, bodily motions, tears, silence, sexual relations, and by a multitude of other means, ritual sacralizes landscapes of various scales and locates them with reference to a human body, a community, or an *axis mundi*. The practice of piety among English Puritans – a ritual venture that encompassed a wide range of everyday life – rearranged, as Max Weber famously argued in his theorizing about the emergence of capitalism, the economic landscape. Hindu ritual bathing in the Ganges River valorizes the lives of beggars and the Hindu spring festival of Holi, in which participants decorate each other with colored powders and waters, features the wholesale ritualized reversal of orders of caste, gender, social status, and age. Ritual war, in the form of Christian crusade or Muslim jihad or Zen-inflected samurai warfare, historically has demonstrated its capability (beyond the obvious) to remake landscape through its reinvention of the relation of spaces to the *axis mundi*, be it Rome, Mecca, or Nippon.

The susceptibility of landscape to change through religious ritual is grounded in respect for the power of the sacred. Religion is predicated on belief in the volatility and danger of the sacred and, equally, on trust in the efficacy of ritual to make possible productive contact with the sacred, to establish territory on which encounter with the sacred can be pursued in relative safety. Places of contact with the sacred – whether that contact be in the form of a felt connection to a deity, saint, prophet, spirit, force, avatar, holy animal, or other divine power – are ritually bounded and constantly policed for territorial leakage. Ritual creates a means by which persons can cross the threshold between profane or secular space and sacred space. Ritual removes persons from their familiar landscapes of status, class, gender, ethnicity, and so forth and prepares them for encounter with awful power. Ritual by the same token reclothes persons in identity when they leave a sacred site, making possible the return to participation in familiar landscapes. The French ethnographer Arnold van Gennep (1960) referred to such rituals as rituals of separation (from the everyday world) and rituals of reaggregation (to the everyday world). What is especially important in the performance of ritual, however, is its capability to define the territory where the sacred dwells. That might be a church or a mosque or synagogue or temple. It might be a shrine or a place associated with an event narrated by scriptures. It might be a human body or a nation. Ritual, as it is performed over and over again, confirms the presence of the sacred in space, and it does so through rehearsal of the boundaries marking that space. Only in this way, through constant attention to the location of the sacred in concrete landscapes, only on this foundation, can religion begin its play with the porosity of borders, the ambiguity of location, and the paradoxical conjoining of distinct, separate territories.

Body

In 1774, St Alphonsus Maria de'Ligouri was at the bedside of the dying Pope Clement XIV. According to accounts, he was also in his cell, in a place four days travel away. The twentieth-century Hindu saint Sathya Sai Baba likewise bilocated, according to reports of his followers. So also, according to religious writers, have Azimia Sufis, Native American shamans, and Tibetan Buddhists. The sixth-century pillar hermit Symeon the Younger, in between demonstrations of raising the dead, was occasionally in two places at the same time. Jesus Christ, in Roman Catholic theology, polylocates every day through his presence in the Eucharist as that ritual is performed simultaneously in a plethora of settings globally dispersed. The fourteenth-century aspiring Christian physicist Antonius Andreae was certain enough of human bilocation that he penned a theologically tinctured tract defending it (Kirschner 1984; Gensler 1999).

Guiding the movement of the human body through territory has long been one of the most important roles of religion. Religion creates body as territory and references the body to its environments. For many religions, the body, first of all, is a composite, a dyad at the least, of a corporeal self and an invisible self or soul. Religion accordingly orients the body not only to the physical world, and all of the landscapes – political, social, ethnic, and so forth – that frame it, but also to the world hidden from ordinary consciousness, the world of angels and demons, gods and spirits. In most cases, religion constructs bodies in a way that situates them at the juncture of two realities, one apparent through the testimony of bodily senses and one made real in visions. This ambiguity is present as well in religious concerns for the body as the repository of the sacred, or as a kind of reservoir of sacred power. Physical space in such instances is defined by the contagion of the sacred as it extends from one body to the next. This can happen while a person is still alive – the healing touch of a shaman or other religious authority (e.g., "the king's touch" in medieval Europe) which extrapolates the miracle-worker's body into the bodies of others – as well as after a person dies and the body, as relic, remains powerfully charged. The relic of a holy person accordingly changes the bodies of those with which it comes into contact in physical space. In Catholicism, the placing of a relic on a table transforms the table into an altar. Buddhist temples and stupas are built around relics believed to imbue the site with sacred power. A hair from the beard of Muhammad at the Topkapi Palace in Istanbul calls for the recitation of the Koran there, uninterrupted and incessant, a ritual mapping of the overlap between the visible and invisible worlds, between the body that was once alive and words spoken by God. Touching the bone of long-dead saint, or even gazing upon it, can heal the most debilitated and sickly bodies of believers.

Relics, which can be purchased in online shops and at shrines, as well as on a thriving black market (since the Middle Ages) represent the commodification of the body, a process advanced at key moments by religion. Religion, for example, cooperated greatly in creating the socio-economic landscape of the slave trade, offering scriptural precedent and theological justifications for human chattel. The pro-slavery arguments of religious leaders in the American South were drenched in religious rhetoric. Old Testament texts narrate numerous accounts of Jewish enslavement of other peoples, including divine collaboration in that. Muhammad took slaves, and Islam institutionalized the practice, with many *hadith* subsequently addressing the sale and treatment of slaves. Such *hadith* undergirded the vast slave trade of Islamic empires, an enterprise that remade the social and economic landscapes of Africa, Asia, and the eastern Mediterranean.

Slavery developed with particular regard to the construction of race, a process significantly aided by religion. Religious scriptures typically make much of differences between holy peoples and unholy peoples. Holy peoples often can be identified by their bodies, which bear the signs of ethnic or racial distinctiveness, and, moreover, are decorated and marked (e.g., circumcision, tattoos, tonsure) in ways that place them within a certain community or demographic. Unholy peoples manifest in their bodies the signs of their unholiness, whether it be behaviorally (e.g., unbridled lust), or through the features of the body, such as skin color, the shapes of the head, nose, lips, height, hair, and so forth. Religious texts are sometimes explicit about such things, and in the hands of ambitious interpreters, those texts can be coaxed into more sweeping and absolute statements about the connection between race and holiness – or, as in the case of pro-slavery argumentation, between unholiness and race. The slave trade in the Americas, which was built on the forced migration of Africans across the Atlantic, was crucial to the colonial and postcolonial social organization of much of the region. The interdependent landscapes of economics, politics, race, and class, as well as human relationships with the physical environment, were grounded in religious understandings of the human body.

Religious ideas about the human body frequently are coordinated with thinking about the nature of society. With anthropologist Mary Douglas (1982), we might think of the two bodies, one social and one the human body, both of which are defined religiously and both of which are subject to the kind of porous-border syndrome that characterizes all religious imagining of place. So, as Douglas argues, in communities where, for example, migration, demographic change, or politics have upset social equilibrium and led to a sense of deterioration of the familiar boundaries of society, that change will be mirrored in thinking about the body. Religious symbology applied to the body will in such instances advance local understanding about what the body is, how it functions, and whether it is healthy or sick, and in social environments where there is a sense of breached boundaries, the body will be imagined as breached as well. Witchcraft fears – which largely amount to concern that demons have crossed the borders of the body, entered it, and begun to take control of it – thus arise in communities undergoing traumatic social change. Exorcisms of demons from the body are performances of anxiety about social problems and wishes for exorcism of the "social body" alongside the human (Douglas 1982). The Communist witch hunts in the United States in the

1950s thus, not surprisingly, took place alongside the rise in popularity of the Reverend Billy Graham, who was outspoken in his claims about the need for persons to be exorcized of evil demons. The visible and invisible worlds of religion in this way converged simultaneously in the social landscape and the body.

Material culture

Religion marks space most effectively through its investments in the built environment. Houses of worship are the most easily recognized markers of sacred space. In their design, materials, siting, and decoration define power, society, economy and commerce, gender, class, aesthetics, and a number of other key structural aspects of culture. Houses of worship are condensed symbols of religious understandings of space and they represent the power of religion in determining culture. A church or temple or mosque is always a center, a point, like the *axis mundi*, from which the directions of the compass are reckoned. It is also a conduit to the invisible world, which typically is located directly above it, in the heavens, and to which the architecture often calls attention, in the form of an elevated nave or sanctuary or ceiling decoration. A house of worship, as repository of the sacred, cannot be entered except with ritual precaution, because the power of the sacred is so overwhelming within it. Threshold ablutions, such as washing in a fountain or sprinkling holy water on one's person, typically fulfill this requirement. By the same token, the power of the sacred that is located within a house of worship must be kept bounded, and so the borders of the site are diligently policed to prevent leakage of the sacred. Leaving a house of worship thus is also a ritual undertaking and often includes a "rite of reaggregation" as persons cross the boundary back into their everyday life. It would be a mistake, however, to assume that the built environment of religion represents merely an attempt to separate the sacred from everyday life. Religious buildings in fact frame ritual performances that reinforce awareness of the regimes of power that organize life outside the cathedral. Seating arrangements, for example, can remind worshippers of class difference (the best seats reserved for those of highest social status), gender distinctions (women worship behind a curtain, from the back rows, or away from the main room in certain synagogues and mosques), and race (African Americans sat in the cold choirs of churches where whites predominated).

Material culture in the form of dress is an important element in the religious demarcation of space. Religious regulations regarding dress include prescriptions for styles of clothing that must be worn to a house of worship or during other religious rituals. More importantly, dress codes grounded in religious views of the world serve as means by which awareness of the difference between the sacred and secular is remembered and enhanced. When a Muslim woman navigates the secular space of the marketplace or street she carries with her, in the form of dress, her boundaries of self, family, worship community, and membership in invisible communions. She marks her body with a scarf or *hijab*, and in so doing, she remembers, and announces, even as she moves about, that her body is located in a certain place. That place is defined by family, fellow believers, religious professionals, and inhabitants of a supernatural domain. There is contiguity with all of those persons just as there is contiguity with the local mosque and Mecca, in spite of the fact that with each step she might put more physical distance between herself and the people and places, smells and sounds, that comprise the familiar environments of her religious life. Religious dress distinguishes her, insulates her, from contact with the profane or secular world. The same is true of monks, in Asian religious traditions as in Western. Devotees of Krishna, known to westerners for their fundraising in public places, do not, in religious terms, inhabit that public place. Their dress (as well as their music, which materially also carves out sacred space for them in various settings) keeps them joined to a holy fellowship of other devotees who might be in that moment at other airports or train stations, and in spite of the appearance of physical distance between them, they remain joined in community. Dress, the borders of the body, defines space by separating the person from the local social environment. At the same time, as is almost always the case in religious figurings of space, it represents believers' trust that their own bodies in fact are not separated from those of other believers, regardless of the physical distance between them. Dress is boundary that confines and sequesters, and simultaneously renders distinctions ambiguous. Dress facilitates bilocation.

Music makes space a religious place, and when it can be heard outside the house of worship it draws that space inside the boundary that distinguishes sacred from secular. Food or the noticeable absence of food also makes space religious. A table set with plates of roasted eggs and chicken wings, celery, horseradish and an apple–nut mixture is a *seder*, a Jewish religious dinner eaten on Passover. The food represents environments significant in Jewish history, and above all the enslavement of the Jews in Egypt and their escape from that. The performance of the *seder* arranges an assortment of places – Egypt, Israel, the antebellum American South, Nazi Germany, contemporary sites of anti-Semitism or genocide – in a pattern of connectedness and places the individual in the midst of that conglomeration. Seder, like much other religious material culture, also plays freely with scale, juxtaposing place defined by family at a table with a global environment of social oppression, while at the same time that it affirms those spaces as mutually constitutive. Material culture in other modes often operates in similar fashion, whether it be the sight of a Kwakwaka'wakw totem pole, the smell of incense in an Orthodox church, the sensation of Ganges River water on the skin, or some other material medium.

Religious practice

Religious space most often is construed as space that bears the markings of institutional religious life: churches and other houses of worship, the presence of persons dressed in religious vestments, the performance of traditional ceremonies, and assorted representations of ecclesiastical power vis-à-vis other institutions. The study of religion and space accordingly has tended in the past to orient itself to spatiality in such settings. As religious studies has coalesced as an area of humanistic inquiry outside the gravitational pull of theological and ecclesiastical discourses, however, it has found other manifestations of religious life to be as rich and as complex, and as powerful in their capability to generate and reinforce meaning, as more well-known formal, institutional forms. Once studied under the rubric of "popular religion" – a term meant to suggest a mode of religious life differentiated from that supervised by formal, elite religion – religious performances in everyday life now are investigated as "lived religion" or "religious practice." This refocusing of research has brought with it new conceptualizations of how religion constructs place and how religion is related to the environments that contextualize persons' lives. The religious landscape in general looks different because of the turn towards lived religion, and the charting of that landscape, while begun in some ways, remains largely an agendum on the horizon of scholarship.

Some of the more promising avenues of analysis of lived religion and space have to do with inquiries into construction of identity through the combining of attributes drawn from a range of religious backgrounds. Lived religion emerges and flourishes largely through the exercise of personal taste, innovation, cultural borrowing, and the blending of seemingly disparate items of religious thought and practice. It manifests as a negotiation between the technologies of piety offered by religious traditions, and the needs and imaginations of individuals. In places where there is considerable plurality in religious life – where there is a fairly broad range of options available to religiously inclined persons – we find in lived religion a reflection of personal attempts to craft religious life in a way that represents a connection to a range of cultural and religious backgrounds. In North America and Western Europe, or other places where migration has been ongoing and profound, lived religion frequently is grounded in the ideas and practices of a Christian denomination, but can include as well components drawn from folk culture, Native religions, Judaism, Buddhism, Islam and indigenous African traditions, as well as from astrology, "vegetarianism," civil religion, and occult healing traditions. The integration of these components relies on personal ingenuity. Once integration is accomplished, personal identity, as reflected in the amalgamation of a number of religio-cultural markers, will reflect the competition and the collaboration between religious worldviews in a particular place. Lived religion of this sort, which tends to be personal or manifest only in small groups, cuts out turf all its own. But because that turf is comprised largely of material imported from a range of places, lived religion implicitly recognizes the authority of other religious places, whether place is defined as community, denomination, or nation.

Sometimes lived religion remains embedded securely in a single tradition, but plays creatively on the borders of that tradition. At other times, it is profoundly syncretistic, and especially so in places where there has been frequent and intense contact between traditions. In Vietnam, Cao Dai began in the early twentieth century as an amalgam of elements drawn from Buddhism, Taoism, Confucianism, Hinduism, Islam, Judaism, Christianity, and indigenous Vietnamese religions. Alongside holy personages associated with these traditions, Cao Dai places as saints others representing an even broader array of backgrounds: Pericles, Julius

Caesar, Joan of Arc, the French novelist Victor Hugo (recalling the French occupation of Indo-China), the sixteenth-century Vietnamese poet Trangh Tinh, the Chinese poet Li Bo, and the Russian writer Leon Tolstoy. The main temple at Tay Ninh, where images of these persons are displayed, incorporates architectural features as diverse as Muslim turrets, Chinese pagodas, European Catholic-style stained-glass windows, interior decoration reminiscent of the creations of Antoni Gaudi, and an artistic rendering of the Eye of Providence, an ancient symbol that can also be seen in the triangular tip of the pyramid on the US paper dollar. Emerging as lived religion in a crossroads place, Cao Dai grew in popularity and soon took on institutional form, adopting a hierarchical ecclesiastical structure modeled on that of the Roman Catholic Church, including the office of Pope. That fact does not mean that Cao Dai is no longer lived religion. Rather, it indicates that human inventiveness is sometimes so effective in crafting a religious landscape through borrowing and recombining that the end product, designed to serve the needs of a single person (originally Ngo Van Chieu), can hold appeal for others whose lives have been shaped by similarly intersecting environments.

The state

Cao Dai, for all of its mingling of cultural traditions and invoking of places around the world, is a strongly nationalistic religion. Indeed, religion has a way of developing in conjunction with national identity so that a national landscape often is contoured in explicitly religious ways. In modernity, and particularly in postcolonial environments, the dividing of space into nations is frequently accomplished through the utilization of religious symbology. In certain parts of the world, Islam, which as a strongly jurisprudential religion is often closely intertwined with state legal systems, identifies a nation as religious space. In certain instances the state is regarded more explicitly and profoundly as a religious entity, as in the case of Japan, where State Shinto (officially until the Japanese surrender in 1945) identified the emperor as a direct descendant of, and high priest in service to, the goddess Amaterasu, who created the land and people of Japan. Emperors, kings, and queens elsewhere, in tribal settings as well as in imperial settings, long have been considered rulers by divine right, and embodiments of good who model that in their virtuous behavior. To declare oneself an adherent of the religion of such a place is to declare oneself a member of the state, and to practice religion is to practice civic virtue. State religion can also incorporate ethnicity, so that to declare oneself a member of an ethnic group can, in some places, locate oneself with respect to national citizenship and religion as well. So, for example, to self-identify as Pashtun is, in most cases, to identify as Muslim and a citizen of Afghanistan, or in the case of a Kalmyk, a Tibetan Buddhist citizen of Kalmykia. International politics, as the early twenty-first century has amply illustrated, accordingly can be heavily laden with religious freight, and particularly so when there is correspondence among race/ethnicity, nationality, and religion.

The emergence in the last 300 years of the secular state in certain ways simplified relations between states, stripping out of negotiations about the occupation of territory emotionally rich issues related to religious practice. This is not to say that religion did not continue to influence the international political landscape, but its influence tended to be oblique rather than direct. Nevertheless the intersection of religion with other environments (social, economic) within a region over time imbues those other environments with traces of the religious worldview, so that even when state religion is deauthorized, other landscapes within the nation can carry forward key aspects of religious tradition. Such has been the case in modern Turkey, where the legacy of the Ottoman *millet* system, which segregated the population (into *millets*) on religious grounds, exercised a determinative influence over Turkish conceptions of nationalism after the formal secularization of the state, with one result being the difficulty of recognizing ethnic difference, and especially the Kurds (Cagaptay 2006).

Time and memory

Space exists in time. One of the most influential theorists of religion, Mircea Eliade (1961), took time and space together in his influential formulations of "sacred time" and "sacred space." While recent research has vigorously pursued the investigation of religion in relation to space, there has been little attempt to join such inquiry to a consideration of the temporal axis. Such an undertaking ought to yield important insights, as for example, in analysis of the cemetery as place constructed out of temporal memes – the past, the present, eternity – that are inseparable from the cultural meanings of cemetery as hallowed ground and religious place. Memory, as one aspect of the temporal axis of analysis, likewise is conjoined with the construction of place. Coming to terms with the past is a part of building a landscape, even if that remembering is willfully traded for forgetting. Imagining the future likewise is crucial to the arrangement of sacred space, to the production of religious environments as part of the larger work of culture. The cutting edge of the study of religion and space should incorporate to an increasing extent analysis of time alongside space. The conceptual apparatus for that project is not yet refined, nor is the electronic technology – particularly in the form of temporally enabled GIS – that might be applied to the investigation of religious landscape in time, diachronically considered. Such an approach would have to be decidedly ecological in its balancing of spatial analysis with attention to timeframe in order to succeed. That is, it will have to recognize that study of religious space in time will have to proceed in the same way that the study of organisms and their natural environments is organized: by principles of interpenetration, symbiosis, cascade, and mutually constitutive realities. Landscape is timescape, something religion has always known. The polylocative is also the polychronic.

11 The cultural production of space in colonial Latin America

From visualizing difference to the circulation of knowledge

Mariselle Meléndez

> There are moreover in that island which I said above was called Hispaniola, fine, high mountains, broad stretches of country, forests, and extremely fruitful field excellently adapted for sowing, grazing, and building dwelling houses. The convenience and superiority of the harbors in this island, and its wealth in rivers, joined with wholesomeness for man, is such as to surpass belief unless one has seen them. The trees, coverage, and fruits of this island are very different from those of Juana. Besides, this Hispaniola is rich in various kinds of spice and in gold and in mines.

These are the words of Christopher Columbus found in a letter addressed to Luis de Santángel (official of the Crown of Aragón) on 15 February 1493 announcing the so-called "discovery" of the "New World." This letter, which circulated around Europe as the only printed official announcement of Columbus' achievements on behalf of the Spanish crown, marked the beginning of the depiction of America to the rest of the world as an entity full of riches now available for material consumption.[1] From this point on, America would be imagined, reinvented, and rewritten as a space full of economic, religious, and cultural possibilities. Space as marked by abundance, marvel, and resemblance was to become an ubiquitous rhetorical theme in narratives as well as visual renditions regarding the arrival of the Spanish to the New World, as demonstrated in the first woodcut prints that accompanied this official letter announcing the event (see Figure 11.1). As Guillaume-Thomas-François Raynal pointed out in his *Philosophical and Political History of the Settlements and Trade of the Europeans in the East and West Indies*,

> No event has been so interesting to mankind in general, and to the inhabitants of Europe in particular, as the discovery of the New World and the passage to India by the Cape of Good Hope. It gave rise to a revolution in the commerce, and in the power of nations; and in the manners, industry, and government of the whole world. At this period, new connections were formed by the inhabitants of the most distant regions, for the supply of wants they had never before experienced.
>
> (1776: I. 1)[2]

Figure 11.1 Woodcut prints from Epistola de insulis in mari Indico nuper inuentes (1494). Note the representation of abundance by the number of buildings, trees, and people. Courtesy of the Rare Book and Manuscript Library, University of Illinois at Urbana-Champaign.

In this global transformation, or as he called it, "revolution" of the world, space came to be a crucial point of departure from which to understand and explain how human and physical geography converged for the sake of Spanish knowledge and power.

Space as a concept and a theoretical discursive tool has long played a major role in colonial studies. Space as a sign of containment, mobility, cultural identity, gender, corporeal expression, mnemonic device, and collective and individual perception has been an indisputably productive way to examine social and gender relations, historical circumstances, and cultural differences. The perception of space becomes part of a process in which an individual is able to define, construct, imagine, or relate to the particular area in which she or he belongs or positions herself or himself.

For scholars in the field of colonial Latin American studies, space has been a useful tool to examine the diverse representations that distinguished the encounter between European and native indigenous societies. Specifically, geography has always been at the center of the many narratives and visual representations published about the "New World." For chroniclers, cosmographers, cartographers, religious and political authorities, America became a geographical entity open to multiple and, many times, contradictory interpretations, as attested by its several names: Indies, New World, and America. As the Mexican scholar Edmundo O'Gorman made clear in 1958, to think geographically about America implied thinking about an idea that deeply transformed the manner in which Europe envisioned the world.[3]

Literary scholars have benefited greatly from the work of critical geographers who have discussed the implications of the role of space as a tool to understand culture, politics, and society in general. Scholars in the field of Latin American colonial studies, for instance, have devoted particular attention to the issue of space as it pertained to the new geographies unveiled by the discovery and colonization. Influenced by the works of J. B. Harley in particular, these studies have paid close attention to the dynamic and critical nature of maps as they are to be considered powerful tools crucial to the construction of that "new" physical reality of what they called America. Geographical narratives and cartography constituted useful tools to examine how spaces as well as places were constructed by becoming part of rhetorical discourses aimed to persuade, manipulate, and impose specific cultural, religious, or political values.[4]

Cultural geography as influenced by poststructuralism impacted the work of literary scholars, who then began to focus on the cultural history of regions as dynamic processes that complicated the colonial exchanges between diverse groups in the New World. Colonial spaces such as the city, the plaza, the church, the convent, and the *cofradías* (brotherhoods) became topics of critical discussions in which asymmetrical relationships of power as well as hybrid forms of expression converged. Other literary critics, influenced by poststructuralist geographers, critically examined the manner in which spaces are visually and discursively constructed as well as epistemologically created in Spanish texts dealing with the colonial Americas.[5] To a lesser extent, literary scholars have explored how indigenous populations, Spanish American Creoles, African populations and other members of *casta* groups, and women also constructed their own territorialities and local spaces as cultural, religious, and political forms of expression. The relevance of this approach relies on offering a broader view of what colonialism as a spatial phenomenon entailed for diverse sectors of the population who occupied what Mary Louise Pratt referred to as "contact zones" (1992: 4).[6] Although these works have not solely focused on spatial issues, they still offer a sense of how racial, social, religious, and gender relations were affected and transformed by the spaces these populations inhabited. In different ways, they point to the fact that spatial relationships are always dynamic and complex, especially when they involve the interaction and clash among different cultures. Space in the colonial context always affects individuals in multidirectional ways, no matter what their social, gender, or racial status.

One area lacking in terms of critical approaches to space in colonial Latin America is the seldom studied eighteenth century, in particular, the influence of the Enlightenment.[7] One can argue that the hybrid character of the discursive production of this period has made them difficult to categorize and fit within canonical literary genres. Even the notion that the Enlightenment made its presence felt in the Spanish territories was not recognized until the pioneering work of Arthur Whitaker (1961), which was later followed by Karen Stolley (1996), Ruth Hill (2000, 2005), Jorge Cañizares-Esguerra (2001), Diana Soto Arango and Miguel Angel Puig-Samper (1995, 2003) and Santiago Castro-Gómez (2005), to name some of the most relevant scholars. Ironically, critical volumes devoted to the Enlightenment as a European movement still fail to include a discussion of the reception of the Enlightenment in Spain as well as in Latin America, offering a very limited picture of the multiple ways in which the ideas of the Enlightenment were understood and transformed in the Hispanic world. As Castro-Gómez argues, in order to understand the manner in which the Enlightenment was read, translated, and enunciated, we need to pay attention to the specificity of location (2005: 15). Space and place came to constitute critical factors in discussing the diverse ideological movements of the Enlightenment in a modernizing world.

C.W. Withers recently suggested that "Rather than being a fixed set of beliefs, the Enlightenment – as a moment and a movement – was a way of thinking critically in and about the world" (2007: 1). Withers adds that the Enlightenment was not exclusively "a historical phenomenon" but also constituted "a geographical one" (2007: 1). It was, as he elaborates, a process that "took place *in* and *over* space – it had a geography, even geographies. It was also *about* space, about the earth, and its geographical variety, and about how that variety–in plants and peoples, cultures and climates-could be put to order" (Withers 2007: 6). Withers's arguments are extremely productive when examining the Enlightenment as a set of dynamic and multiple processes depending on their particular locations of enunciation. In an age when the circulation of knowledge from a transnational, continental, and global perspective reached new dimensions in terms of how physical spaces were conceived as material objects of production and consumption, it makes sense to think about the Enlightenment in terms of space.

In this essay, I examine how a group of Peruvian intellectuals reimagined their patriotic space in terms of particular ideas of the Enlightenment that circulated in colonial Latin America. My discussion centers on several news articles published in the Peruvian newspaper the *Mercurio peruano* in 1791–5 by a group of Creoles who acted as editors of this weekly text.[8] If colonial Latin America was widely read and understood in spatial terms from the European centers at the time, what I would like to propose here is a different reading of that same geography that

Europeans observe, and were trying to explain and categorize. I would like to focus instead on how Creoles themselves read their own spaces in dialogue with a national and international public. I argue that geography in particular was intrinsically connected to the manner in which Latin Americans understood and discursively produced images of their own territories for a European public. For them, space was conceived in patriotic and utilitarian terms, and was also a source of national prestige, demonstrating the intrinsic relationship which existed between space and power when it came to the articulation of social and cultural differences in a colonial setting.

Visualizing local spaces as signs of prestige: the case of the *Mercurio peruano*, 1791–1795

In 1790, a prospectus announcing the publication of the *Mercurio peruano* was released to the public explaining to the readers the reasons behind the creation of the newspaper and its ultimate goals.[9] A crucial goal for the founders of the newspaper was to ignore news from other parts of the world and instead focus on their own homeland: "What interests us the most is what happens in our Nation instead of what interests the Canadian, the Laplander or the Muslim."[10] As "lovers of public enlightenment," they paid special attention to what they deemed made their country unique. Disseminating that knowledge was key in their intellectual endeavors.

The editor pointed out that, when it came to Peru, there was a lack of news that corresponded to the greatness of a country "so favored by Nature due to its temperate Climate, and the abundance and richness of its Soil."[11] For the editor, it was a shame that a country so well endowed with natural resources was scarcely known to the rest of the world. It was precisely this lack of accurate information about their kingdom that motivated the founders to create the newspaper.[12] Peru as a geographical entity became the center of their discussion, emphasizing the relevance of their country's natural resources and economic potential.

The author of the prospectus emphasizes the need to inform readers about Peru's history, its people, monuments, commerce, ports, agriculture, mining industry, fishing, and other aspects of natural history to better understand what made Peru such a special territory. Each piece of published news would reiterate why Peru had to be considered an enlightened nation. The challenge posed to themselves as "Lovers of happiness and of public enlightenment" (Amantes de la felicidad y de la ilustración pública) was to put Peru in a place of international relevance by making an international public aware of the material as well as intellectual richness of their country. For them, space became a productive tool to disseminate knowledge of their country, envisioning it, as Lefebvre suggests, as a "social reality" and "a set of relations and forms" (1998: 116).

The first news article published in the *Mercurio peruano* encapsulated the image of Peru that the editors wanted to share with the rest of the world. The article, entitled "Idea general del Perú" (A General Idea of Peru), aimed to offer a succinct but accurate picture of those elements that made Peru a distinctive space.[13] As the article points out, "the principal objective of this Newspaper as it was mentioned in its Prospectus, was to make this country that we live in better known; this country against which foreign Authors have published so many false statements."[14] The editors were referring to what Antonello Gerbi called "the dispute of the New World," sparked by the writings of Europeans such as Denis Diderot, Cornelius de Pauw, George-Louis Leclerc de Buffon, William Robertson, Amédee François Frezier, and Guillaume-Thomas-François Raynal, who emphasized "the 'weakness' or 'immaturity' of the Americas" when it came to people, fauna, flora, and geography in general (1993: 3). These views, as Gerbi adds, came from "the tendency to interpret the organic link between the living and the natural, the creature and its environment, as a fixed, necessary and causal relationship" (1993: 29). For these authors, America as a continent was characterized by a vast but poor and hostile nature, less stable and more decadent species, and decrepit, lazy, and immature people. Such statements from some European authors who had never visited America prompted a series of responses by Creole and *Mestizo* intellectuals such as Francisco Xavier Clavigero, Juan de Velasco, Eugenio de Santa Cruz y Espejo, and many contributors to Spanish American newspapers questioning the generalizations made with regard to Latin America. Within this context of this dispute, articles such as "General Idea of Peru" emerged.[15]

One of the major problems that the editors of the *Mercurio peruano* found with regard to the European versions of the history of Peru was the fact that they were guided by particular national agendas and were quite often ignorant about these territories. In a viceroyalty as vast as Peru, it was impossible to find accurate historiographical works that could comprehend the vastness and distinctiveness of its territories.[16] It was not, according to the editors, until the publication of Jorge Juan and Antonio de Ulloa's *Relación histórica de un viaje a la América Meridional* (1748) that a more reliable history of Peru from an European perspective appeared. Jorge Juan and Antonio de Ulloa were part of the scientific expedition that took place between the years 1735 and 1744 authorized by Philip V to measure the size and shape of the earth at the equator and to determine its oblateness.[17] This expedition constituted part of a modern scientific approach to study with precision not only the shape of the earth but also the nature of these territories with regard to fauna, flora, geography, commerce, and agriculture, among other matters. However, Spanish and French scientists' views of Peru did not necessarily coincide with the way in which Peruvians perceived their own country. The rationalization of that space in terms of its utility for mercantilist purposes, for economic and social progress, for the globalization of material goods, and for the development of modern scientific achievements was a key factor in the manner in which American territories were envisioned by Spanish Americans as well as Europeans.

The main problem Peruvian Creoles perceived about foreigners' depictions of their land was that they relied on overt generalizations. It was mostly the specificity of place which the editors wished to emphasize in their general history of

Peru. They believed it necessary to view their country through the eyes of those who have been immersed in its history. Indeed the editors pointed out that the histories, geographical treaties and compendiums, letters, and reflections that European writers had published offered a distorted version of Peru to the extent that it seemed "a complete different country from the one that their practical knowledge show them."[18] They found it then necessary to present the public "exact news accounts" of their own country, differing greatly from the news published by Europeans.

Their first task was to make clear the dimensions of the viceroyalty of Peru, emphasizing the vast territories comprising it. It included a diverse geography and topography, including arid hills, large extensions of sand, voluminous lagoons, villages and cities surrounded by pleasant valleys, amazing mountain ranges, and a variety of climates. The kingdom was also endowed with a geography conducive to the extraction of minerals and the business of agriculture. With regard to the mining industry, the article emphasized that it was "the main and only source of riches of Peru" albeit this industry had not been exploited enough because the mine owners had not received the appropriate economic incentives from the Spanish government.[19] They argued that the mines of Gualgayoc and Pasco counted for half of the silver that was produced in Peru (Figure 11.2). With regard to agriculture, the editors complained that it had not been developed well. The variety of climates and topography endowed Peru with the possibility to cultivate a myriad of agricultural products to the extent that they would not need to import any item from foreign countries. However, the poor development of adequate transportation routes had made the circulation and expansion of agricultural exports almost impossible. In sum, the territories encompassing the viceroyalty possessed the diversity and abundance needed to elevate Peru into an influential position in the global market; an overall lack of successful development and investment had prevented its full economic potential. In the end, Peru's natural history was an illustration of "the wonders" that made the kingdom fecund and unique. This fact, combined with the idea that "the Enlightenment was general in all Peru," producing studious, sharp, and well-prepared citizens, made Peru an ideal place for economic development.

In an effort to demonstrate in more detail the potential of the country to be an economic power, the editors published another article entitled "Introduction to a Scientific Description of the Plants of Peru" (Introducción a la descripción científica de las plantas del Perú). This article served as an example of the manner in which the editors visualized their country and discursively produced it for the rest of the world. The editors complained that Peru had remained an unknown commodity when it came to the richness of its flora. The variety of plants that indigenous communities in the past used for medical and agricultural purposes and "so many utilities" (muchísimas utilidades) were barely known to Europeans and even to Peruvians themselves. Even the European scientific expeditions that reached Peru prior to 1778 were not able to capture the abundance and variety of plants or to order them in an intelligible manner (Figure 11.3).[20] It was not until

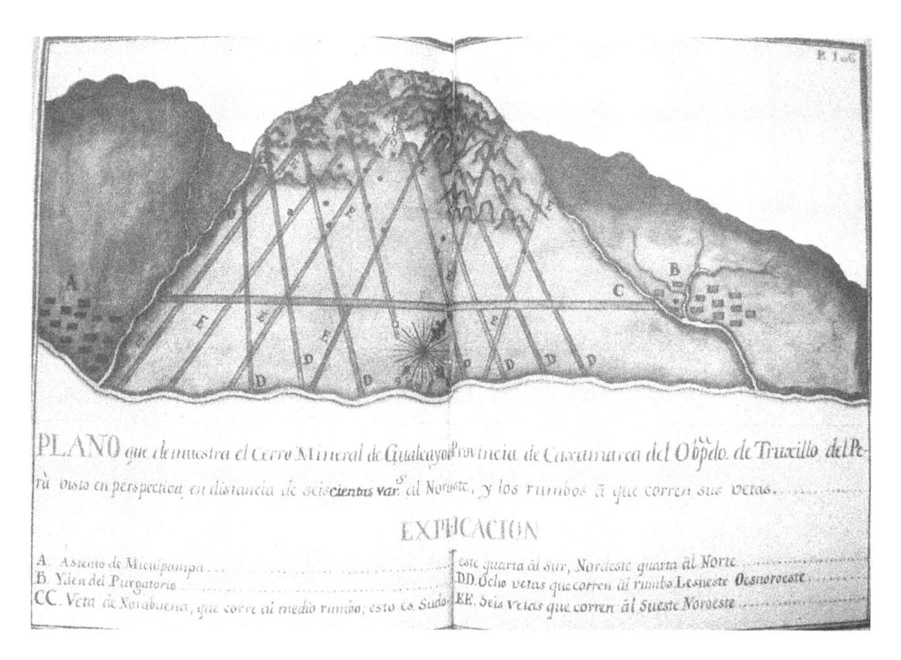

Figure 11.2 Mine of Gualgayoc as depicted by the bishop of Trujillo del Peru in his collection of watercolor illustrations of his diocese, 1782–5. Courtesy of Biblioteca del Real Palacio de Madrid.

1778, according to them, that the expedition led by the botanist Joseph Dombey, Hipólito Ruiz, and Joseph Pabón revealed this issue to the public.[21] The editors of the *Mercurio peruano,* aware of the intentions of the Spanish government to publish the findings on Peruvian flora, declared that the publication of such a document would represent an "eternal monument of wisdom and magnificence," an "opulent treasure of the vegetal kingdom" and "the most authentic testimony" about Peru's natural richness not only for the abundance of precious metals but also for the great variety of its "exquisite plants" (1791: 243. 75–6). According to them, it was imperative to share with the public the importance of this flora as it pertained to "the common utility" (utilidad común) of the country. Peru, as a space endowed with rich flora, constituted an ideal place for a productive economy based on its natural resources. By dividing the flora by class, categories, gender, families, varieties, and individuals, people were able to better order and control those natural resources. The article compared the vegetable kingdom to "a country" or to "a numerous army" that when organized was able to bring forth power and success. Just as a city needed to be organized in plazas, streets, and towns and people needed to be categorized by their social class, it was important also to order and categorize Peru's flora to better take advantage of its commercial possibilities. By following Linneaus's system of classification, Peruvians could understand how useful it could be for the arts, sciences, and most importantly, for

Figure 11.3 "Ficoides peruvina" and "Elichrysum Americanum" as portrayed by Louis Feuillée in his expedition to South America, 1709–11. Courtesy of the Rare Book and Manuscript Library, University of Illinois at Urbana-Champaign.

agriculture. As the editors reiterated, with an enlightened view of Peru's botanical organization, agriculture could be improved in order "to leave behind the miserable desertion" that had been found to date.[22] They believed commerce would grow via cultivation of more diverse agricultural goods. Also, the field of science would improve by making Europeans aware of medicinal plants that were not found on their continent yet were beneficial for scientific progress. The categorization of plants would also help those less knowledgeable in botany to avoid confusion in the use of erroneous plants for specific medical purposes, as was common at that time. They believed the study of botany represented a useful tool "to the benefit, enlightenment and honest pleasure of Men" (á la comodidad, ilustración y honesto placer del Hombre: 2/44. 85).

These articles demonstrate how this Peruvian newspaper served as a privileged locus of enunciation from which to disseminate knowledge about the physical geography of the country. For the editors of the newspaper, geographical knowledge brought recognition to the greatness of their country in terms of creating awareness of the natural resources waiting to be incorporated into the global economy. However, it was important for Peruvians, as well as the Spanish government, to understand Peru's full economic potential in order to take advantage of these resources. Their position corroborated that of many eighteenth-century intellectuals who, as Withers observes,

> understood their world to be changing as a fact of geography, and as the result of processes of geographical inquiry-in the shape and dimension of continents . . . in the types of human cultures making up mankind, in the reason plants, animals, and human races were located as they were.
>
> (2007: 5)

In the case of the editors of the *Mercurio peruano*, Peru was perceived as a privileged space still unknown to many Peruvians themselves and Europeans. It was their duty as "lovers of the country" to place Peru in the epistemological map of the Enlightenment by exposing its capacity to compete in a modern economy. Space as a useful tool to rethink their country's potential constituted a crucial element of the patriotic project in the *Mercurio peruano.*

This patriotic project was also reflected in their representations of urban space. The editors of the newspaper perceived Lima, capital of the viceroyalty, as an enlightened city populated in its majority by illustrious men and women who were highly literate.[23] Lima was the source of their "patriotic love", a sentiment, that was displayed in the buildings and monuments that were located at the center of the city. In a lengthy article describing the Plaza Mayor entitled "Description of the Famous Fountain seen in the Plaza Mayor of this City of Lima," the editors described in detail the fountain, which aside from decorating and enlivening the city, also constituted a useful object for the city and its citizens.[24] They argued that the fountain benefited the city by requiring the construction of efficient aqueducts, which enabled water to be made available in places that had been impossible to serve in earlier times. The construction of such fountains partly contributed "to perfect the science of the movement of waters preserving a portion of the water in a determinate place" (1792: 4/116. 100). The editors encouraged people to observe, examine, and analyze these monuments to better understand their multiple useful values that went beyond "a trivial" occupation of space. However, they acknowledged that, since antiquity, fountains also had been considered signs of opulence by nations that competed for social prestige and recognition. Lima's fountain was no exception. Made of copper, bronze, and masonry boasting exquisite moldings and intricate carvings, as well as beautiful glazed tiles; the fountain stood at an imposing height, and had 46 spouts.

Lima's fountain had nothing to envy to those erected in Rome, Luxembourg, Versailles, or Madrid. As they noted, the fountain was proof of "magnificence and good architectonic taste" (1792: 4/115. 9). Nevertheless, the editors made sure to reiterate to their readers that all citizens should also remember the fountain's value when it came to "public utility" (utilidad pública) (1792: 4/116. 101). For the editors, Lima's plaza and other urban spaces such as cafés, anatomical amphitheaters, the university, and churches and convents constituted signs of Lima's prestige and proof that it was indeed an enlightened city. At a time when Bourbon reforms on urban policies were changing the structure of urban areas, these "lovers of the country" set out to publicize what they considered the value of monuments, buildings, and places in the colonial capital.[25]

Nevertheless, it is important to note that rural spaces were also a high priority for these editors as they aimed to unveil the relevance of these territories as part of a more inclusive economy. Interior areas of the viceroyalty were a major concern for Spanish authorities, as illegal trades by local and foreign parties as well as a lack of cultural and political integration were the norm. The editors of the *Mercurio peruano* reiterated that, due to the diverse topography and climate, these territories abounded in potential riches, ranging from a variety of fruits, vegetables, herbal plants, and minerals, to the possibility of developing additional manufactured items based on the production of a variety of consumer goods.

One point of contention was the idea held by many European travelers that indigenous people who inhabited those territories were not fully integrated into the colonial economy. Since it was believed that these people were occupying lands full of possibilities but hardly exploited, what was thought to be needed was a precise knowledge of the territories, an incorporation of the population into the economy, and a good administration of those resources. As Antonio de Ulloa and Jorge Juan made clear to the Spanish king in their private report known as *Noticias secretas de América* (*Secret Report about America*):

> All these things that Peru produces, and many other particulars found on those vast kingdoms and countries, which news are ignored because of lack of attention, could represent riches for any other nation who knew how to grant them the estimation that they deserved . . . nevertheless we still do not know how to take advantage for our own benefit, and this constitutes the essential reason why we do not show the riches that are produced in our Indies.[26]

As expected by the Bourbon administration, the useful knowledge of these interior lands would help "to consolidate political control over some of those strategic frontiers, secure them from Indian raiders and foreign interlopers, and make them more productive" (Weber 2005: 5).

The editors of the Peruvian newspaper also considered religion a valuable tool for the integration of these interior lands and the indigenous habitants who populated them into the country's economy. In a lengthy article written by Father Manuel Sobreviela and published in the *Mercurio peruano* in October 1791, the impetus for integrating Indians into the economy became apparent.[27] This article was accompanied by a map that accounted for lesser-known interior territories within the viceroyalty. The article offered relevant information about the entrance of the Franciscans into the remote areas of the Peruvian Andes and their involvement in the religious conversion and social integration of the indigenous population that inhabited those areas. At a time in which the Bourbon administration was reducing the power of religious orders by privileging secular clergy to occupy ecclesiastical positions, this article offered not only a claim that religion still constituted an important tool for the integration of remote populations into the colonial system but also that this particular religious order was the most prepared, as it was able to successfully occupy the difficult and still unknown fringes of the viceroyalty.[28] As Father Manuel Sobreviela explained, "Since 1637 to the present, fifty-four religious men were killed in the mountains of Peru at the hands of infidel Indians," showing the dangers and difficulties that those who had survived had to endure (1791: 3/80. 92).[29]

One aspect that Sobreviela, as well as the editors of the newspaper, highlighted as a justification for publishing the article was the inclusion of the first map ever available of these remote areas of the viceroyalty. For Sobreviela, the importance of publishing the map resided in the fact that it visually conveyed to colonial authorities the paths needed to be taken through the mountains of these regions so they could reach – as the Franciscan had done – "the countless barbaric Nations" (innumerables Naciones barbaras) that inhabited those territories (1791: 3/80. 92). Conversely, for the editors of the newspaper, the map served as proof that natural and human resources in remote territories could potentially be controlled. The paths illustrated in the map also represented the troubles and difficulties that the Franciscans encountered, leaving "their blood" as a mark at every step of the way. Although the map was unable to visually reproduce that blood, it did show the presence of viable mission communities in some of the most remote regions of the viceroyalty (Figure 11.4).

The map shows a detailed and comprehensive view of the territories comprising areas located in the Andean mountains, reiterating how these lands were being religiously conquered.[30] The depiction of the indigenous family on the right side of the map offered a visual reminder of the heathen Indians (indios gentiles) who inhabited these areas, as the words in capital letters indicate at the center of the map: "Habitadas de Gentiles. Pampas del Sacramento." The family is depicted holding three artifacts that denoted their barbaric status. The man is holding an arrow and wearing his original clothing, which displayed no signs of European influence. Next to him, the naked child at the center is holding a bird by the neck, signifying his/her domination over nature. The woman is half naked, and ready to eat the bird they apparently had just hunted. The visual depiction of the indigenous family reinforced the need to integrate these inhabitants into civilization through the vehicle of religion. For the editors of the newspaper, religion and the Enlightenment were not mutually exclusive as the former served as a tool to educate, give order, manage society, and control spaces.[31]

The material structure of the map denoted the task already achieved but also the one that still remained, as it differentiated the lands already converted and controlled from those still populated by gentiles. The human space was divided into three groups: (1) "land of Christian Indians," (2) "lands of converted

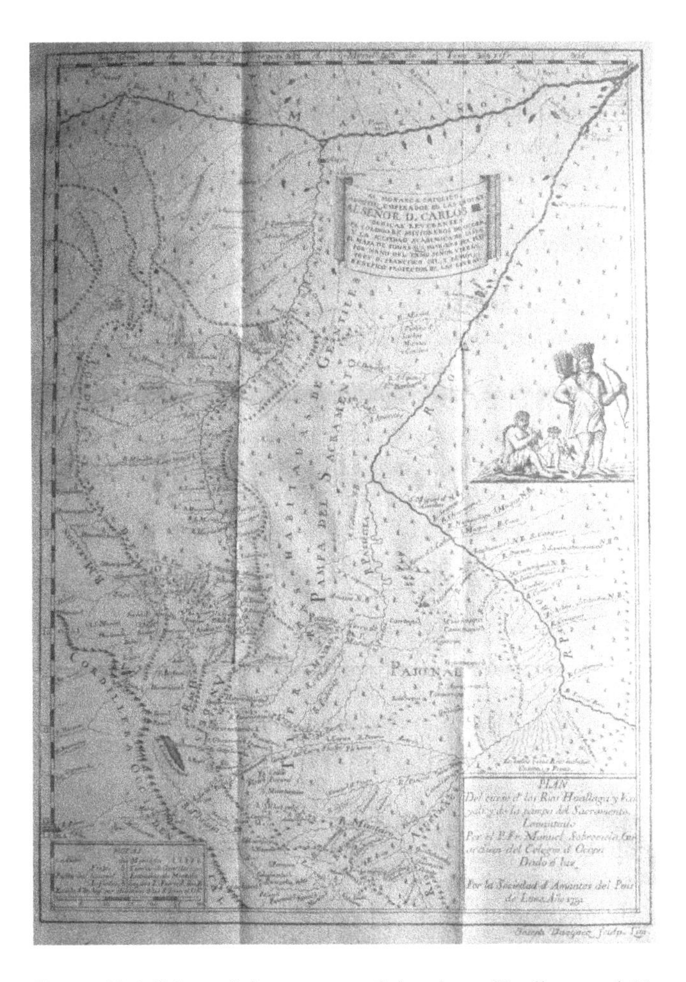

Figure 11.4 "Map of the courses of the rivers Huallaga and Ucayali in the Pampa of Sacramento," composed by Fr. Manuel Sobreviela and engraved by Joseph Vazquez in Lima, 1791. Courtesy of the Rare Book and Manuscript Library, University of Illinois at Urbana-Champaign.

Indians," and (3) "land of heathen Indians" (see notes on the bottom left). For the first two groups, the cross served as a reminder of religious success. In addition, topography went hand in hand with the demography of the territories. The visual icon used to depict mountains stressed the idea of difficult and dangerous territories they had to confront in order to integrate these Indians into civil society. The map emphasizes that the task was possible if headed by the appropriate persons, suggesting Franciscan missionaries were able to master those lands. By providing the Spanish king with a map of these remote territories never seen before by

Europeans, Sobreviela and the editors of the newspapers unveiled for the king and the public the idea that what had been considered a no man's land or unconquerable space was not necessarily true. The long list of rivers and their locations as well as the lagoons surrounding these areas conveyed the possibility of economic remunerations if the lands were appropriately managed and developed. The detailed codification of human and geographical knowledge worked as a rhetorical tool to communicate to Spanish authorities and the rest of the world with access to the newspaper[32] how these lands as well as inhabitants were not totally lost to the kingdom but instead could represent a viable source of economic exploitation. For the editors of the newspaper, in these zones, religion seemed to be the most useful tool to make this process possible.

The fact that the map was addressed to Charles III emphasized the respective political agendas of the Franciscan missionaries as well as the members of the Academic Society of the Lovers of the Country. The materialization of space for religious and economic purposes aimed to serve as incentives for the Spanish government to allow the existence of Franciscan missions in those areas. The map also demonstrated that these indigenous populations were not as impossible to convert and civilize as many European thinkers like Robertson, Raynal, and De Pauw thought at the time. For the editors of the newspaper, it was also important to emphasize the location of the Huallaga and Ucayali rivers as viable systems of transportation. Furthermore, the Pampa de Sacramento, or the area between the two rivers, constituted very fertile and diverse zones. The only obstacle to the development of this unique geographical region was a means of totally integrating the indigenous population of the east central Andes into the kingdom's economy.

The map served as a tool to gain better knowledge of these regions and to exploit their potential. The publication of the map in the Peruvian newspaper constituted part of that project that, as Christian Jacob suggests, characterized any construction of a map: the aim of "organizing an codifying knowledge" by symbolically appropriating space for the sake of particular religious, and economic agendas (2002: p. xix). In this case, the map constituted as Jacob also suggests, a "rational construction" subject to interpretations due to its materiality or "graphic characteristics" (2006: 2). In Sobreviela's map, human and physical geography were intertwined, as the caption and the visual representation depicted. The Enlightenment emphasis on geographical knowledge for the purpose of understanding the usefulness of space for economic gains played a major role in the editors' decision to publish the map, which, they thought, was a way to facilitate potential control of that space. The map constituted a visual discourse that facilitated an epistemology of local spaces for religious purposes in the case of the Franciscan missionaries, as well as for economic and patriotic purposes in the case of the editors of the *Mercurio peruano*. In their effort to insert Peru into a more visible position, the editors found in this unpublished map a means to unveil the interior of the Peruvian viceroyalty to the Spanish administration and the rest of the world. As Jacob observes, mapping in this sense becomes

> a speculative process in which the graphic mechanism attests to the symbolic violence inherent in every model, that is, to the transformation of real

space into a figure ruled by laws of reason and abstraction, of the conquering appropriation of reality by means of its simulacrum.

(2006: 23)

Concluding remarks

The *Mercurio peruano* reveals how space constituted a theoretical ground from which to think about issues that the Creole editors deemed important with regard to the image of Peru they wanted to share with their fellow citizens and the rest of the world. Their *patria* (homeland) was conceived in spatial terms within the framework of ideas that circulated at the time. Nevertheless, their reading of the Enlightenment was transformed by the manner in which they articulated the vision of their country. The stress on economic progress by maximizing resources, the rationalization of space as a means to take advantage of natural capital, the production of knowledge grounded in the utility of space, and the emphasis on putting human and physical geography in order to maximize their economic potential constituted parts of an Enlightenment view of the world that the editors adapted to their own country and needs. It was paramount to demonstrate to the rest of the world that Peru possessed not only the human capacity to make progress in their own country a true reality, but also that Peruvian territories in their topographical and climatic diversity constituted fertile lands, with great a abundance of mineral and agricultural goods and varied flora. In sum, Peru possessed all the natural assets needed that, if well administered and developed, would transform the country into a major player within the modern global economy.

Space has always been a productive concept to examine the process of colonization in Latin America. Space, however, does not represent a new critical tool to study the multiple ways in which Latin America was conceived from the fifteenth to the eighteenth centuries. Since the arrival of the Europeans in the Americas, and even in precontact societies, space has been a useful and critical tool to think about the transformations imposed upon the world, daily experiences, the manner in which human and physical geography collided, and religious, cultural, economic, and political interests. In the eighteenth century in particular, Enlightenment ideology as developed in science, economy, and politics made space a more prevalent theoretical point of reference to observe, learn, debate, categorize, and think about the world. The sites of knowledge as well as the local circumstances made the Enlightenment a dynamic intellectual and geographical movement from which to approach the world, always guided by specific situational interests. Indeed, the emphasis on geographical knowledge at the time, as Withers suggests, was possible through activities such as "encountering and imagining, mapping and inscribing, envisioning and publicizing" (2007: 88). For the editors of the *Mercurio peruano*, as well as for other Spanish American intellectuals of the time, space played a key role in the manner they envisioned their countries within a national, transnational and global perspective. The hermeneutics of space took a variety of forms in eighteenth-century Latin America, many of them still seldom explored. The need to make their own space visible and known to

Europeans who, living at a great distance and without traveling to those territories, made false generalizations about the Americas, constituted an incentive behind many of their debates. After all, as the editors of *Mercurio peruano* argued in their "Introduction to the Scientific Description of the Plants of Peru," it was "a speculative and exact eye" that could register and understand the totality of their country; a mission that they executed from a privileged and unique discursive space (1791: 2/43. 74).

Notes

My deepest thanks to my research assistant, Marcos Campillo-Fenoll, for his help gathering all the visual material as well as some primary sources included in this article.

1 As Margarita Zamora argues, this letter published with a retroactive date was a rewriting of the original letter with which Columbus originally had addressed the King and Queen dated 4 March 1493 (Zamora 1993: 9). For more information on the discrepancies between both letters and the reasons why the "Letter to Luis de Santángel" became the official document of the announcement see Zamora (1993: 9–20).

2 The original edition was published in 1774, in La Haye, France under the title *Histoire philosophique et politique des établissemens & du commerce des Européens dans les deux Indes*.

3 For more information on the invention of America as a geographical entity see O'Gorman (1984). For a more recent discussion on this issue; see Mignolo (2005).

4 The works of Zamora (1993), Padrón (2004) and Wey Gómez (forthcoming) have benefited greatly from studies on cartography by focusing on how spaces are transformed and manipulated for ideological and imperial purposes. It is not my intention to offer a bibliographical summary of these critical works but rather to provide the reader a general idea of how literary critics dealing with the early colonial period have approached the phenomenon of space. For a detailed discussion on some of these works, see Arias and Meléndez (2002: 13–17). It is also important to clarify that my discussion does not include the important contributions of historians who have also worked with spatial issues on colonial Spanish America; I am alluding only to literary scholars.

5 See Mignolo (1995, 2000), Rabasa (1993, 2000), Arias and Meléndez (2002), Pratt (1992), and Verdesio (2001) on this topic.

6 For example, critical works dealing with the convent as a space deeply marked by social and racial differences or envisioned as a space of gender autonomy (Arenal and Shlau 1989; Ibsen 1999; Myers 2003; Eich 2004; McKnight 1997), or studies on indigenous conceptualizations of space as crucial in processes of identity construction (Adorno 1988; Castro Klarén and Millones 1990; Mignolo and Boone 1994; López Baralt, Velazco 2003), or studies focusing on the manner in which the black population in colonial Spanish America was affected by the spatial control of their bodies (Eich 2004; Meléndez 2006) constitute some of the studies that have viewed and studied particular spaces and places of location as productive critical tools to understand the many forms in which colonialism took place in colonial Spanish America.

7 For an excellent discussion on the particularities which have made the eighteenth century "a literary no-man's land" as pertains to colonial Spanish America, see Stolley (1996: 336–74).

8 The term "Spanish American Creoles" refers to individuals born in the Americas from Spanish descent.

9 The *Mercurio peruano* was founded by the Sociedad Académica de Amantes del País (Academic Society of Lovers of the Country), a group of young intellectual Creoles mainly from Lima, whose expertise ranged from medicine to commerce, science, geography, religion, literature, and law. The founding members of the Academic Society became the founders of *Mercurio peruano*, and according to the anonymous author were Hermágoras (José L. Egaña and president of the Academic Society), Aristio (José Hipólito Unanue, the Secretary), Hesperióphilo (José Rossi y Rubí) and Homótimo (Demetrio Guasque). Cefalio (José Baquíjano), Théaganes (Tomás Mendes) and Archidamo (Diego Cisneros) were also three important members of the Society. According to Manuel de Mendiburu, of the 30 academics who belonged to the Academic Society, 21 were from Lima (1890: 8. 158). All members of the Academic Society became active contributors of the newspaper. For biographical information on some of these contributors, see Mendiburu (1874). For more information on the archival project of the *Mercurio peruano* see Meléndez (2006) and Clément (1997).

10 The Spanish phrase reads "que mas nos interesa el saber lo que pasa en nuestra Nacion, que lo que ocupa al Canadense, al Lapon, ò al musulmano" (*Mercurio peruano* 1791: 1/1). For the Spanish citations, I will follow the original orthography from the facsimile edition. When quoting from the newspaper, I will cite the volume, number and year of the publication as well as page number. The prospectus was not numbered. All translations from Spanish to English are mine unless otherwise specified. I have tried to maintain the nature of the Spanish discourse as much as possible, sacrificing at times more common English grammar and expressions.

11 The Spanish original says "[el Reyno peruano] tan favorecido de la naturaleza en la benignidad del Clima, en la opulencia del Suelo" (1791: 1/1).

12 The editors use the name Peru to refer to the viceroyalty of Peru, which in 1791 comprised what is today considered today as Peru, but up to 1717 included all of South America except parts of Venezuela and Brazil. They also referred to it as the kingdom of Peru.

13 Although some critics name José Rossi y Rubí as the author of this news article, the article does not list his name and it is written in a collective voice referring to the editors and founders of the newspaper itself.

14 "El principal objeto de este papel Periodico, segun el anuncio que anticipo en su Prospecto, es hacer mas conocido el Pais que habitamos, este Pais contra el qual los Autores extrangeros han publicado tantos paralogismos" (1791: I/1. 1).

15 For a detailed discussion of this dispute see Gerbi (1993: 3–324) and Cañizares-Esguerra (2001: 1–129). See also S. Arias in this collection.

16 In an attempt to centralize the government to better administer these territories, two new viceroyalties were founded. The establishment of the Viceroyalty of Nueva Granada incorporated in 1717 and again in 1739 encompassed what is today Venezuela, Colombia, Ecuador, and Panama. In 1776 the Viceroyalty of La Plata was established, having jurisdiction over Chile, Argentina, Uruguay, and Paraguay.

17 This expedition was led by members of the French Academy of Sciences including the geographers and mathematicians Louis Godin, Charles de La Condamine, Pierre Bouguer, and other French naturalists. The expedition led to the publication of several accounts of the scientific adventure from the Spanish perspective (Antonio de Ulloa and Jorge Juan) and the French perspective (Charles de La Condamine and Pierre Bouguer). For more information on this expedition see Pratt (1992: 16–23).

18 The original quote reads "parece un pais enteramente distinto del que nos demuestra el conocimiento practico" (1791: 1/1. 1).

19 The quote reads, "La Mineria es el principal y tal vez el unico manantial de las riquezas del Peru" (1791: 1/1. 4).

20 The expeditions to which the editors referred were led by Louis Feuillée, which took place between the years 1709–11 and whose observations were later published under the title *Journal des observations physiques, mathematiques et botaniques, faites par ordre du roi sur les côtes orientales de l'Amerique méridionale, & aux Indes Occidentales. Et dans un autre voïage faite par le même ordre á la Nouvelle Espagne, & aux illes de l'Amerique* (1725), and the one in 1736 already mentioned in no 17.

21 The expedition was organized by the French government with the aim to find and collect plants that could be cultivated in France. The Spanish crown gave permission with the condition that two of its botanists be appointed as part of the expedition and that their findings would be published first by the Spanish crown. The Spanish government was interested in this type of botanical inventory as it was a way to learn more about Peru's natural resources and the possibility of its exploitation. Dombey's drawings were seized first by British privateers as well as Spanish authorities, who prevented Dombey from sending the drawings to France. The drawings were later utilized by Ruiz and Pabón to write their *Florae Peruvianae, et Chilensis* (1794). Ruiz also published *Relación histórica del viage que hizo a los Reinos del Perú y Chile* (1777–8).

22 The original quote reads "la Agricultura podra mejorarse con las luces que vamos á esparcir sobre ella, y salir del miserable abandono en que se halla" (2/44. 83).

23 Of course, the editors did not hesitate to comment that cases of social deviance and disorder in the city had to do with repugnant celebrations held by other sectors of the population such as blacks and individuals of African descent. On this issue see Meléndez (2006: 212–19).

24 The Spanish title is "Descripcion de la famosa fuente que se ve en la Plaza mayor de esta Ciudad de Lima." The article was published in two different issues of the newspaper.

25 For a discussion on some of the urban policies established by the Bourbons in the second half of the eighteenth century as a result of their desire to centralize colonial administration and as a consequence of the terrible earthquake that hit Lima in 1746, see Walker (2003: 54–8).

26 The Spanish quote reads "Todas estas cosas que el Peru produce, y otras muchas que habrá particulares en aquellos dilatados reynos y payses, cuyas notician se ignoran por falta de aplicacion, serian riquezas bastante para otra nacion que supiese darles la estimacion que merecen . . . sino que aun no sabemos aprovecharnos de ellas para nuestro propio uso, y esta es la causa esencial de que entre nosotros no luzcan las riquezas que producen nuestras Indias" (1983: 601). *Noticias secretas de América* constituted a private report about the real economic, political, and social status of the viceroyalty of Peru as a result of the observations taken by Jorge Juan and Antonio de Ulloa during their participation in La Condamine's expedition. The report offered an image of the problems confronted in the viceroyalty with regard to the treatment of the indigenous population, contraband, foreigner incursion in the interior lands, lack of colonial administration, legal corruption, abuses by ecclesiastical authorities, and inappropriate use of its rich natural resources. David Barry in 1826 published the report for the first time in England.

27 Father Manuel Sobreviela belonged to the order of the Franciscans and was a preacher and guardian of the College of Santa Rosa de Ocopa, a small town in the foothills of the Andes where he arrived in 1785. He was from Aragon, Spain. The Franciscans arrived in this particular region in 1725. The title of the article is "Varias noticias interesantes de las entradas de los Religiosos de mi Padre San Francisco han hecho á

las Montañas del Perú, desde cada uno de los partidos confinantes en la Cordillera de los Andes para mayor esclarecimiento del Mapa que se da á la luz sobre el curso de los Rios *Huallaga* y *Ucayali*." For a complete biography on Father Sobreviela see M. Mendiburu, *Diccionario histórico biográfico del Perú*, (1887: 7. 351–5).

28 The goal of the Spanish crown was to secularize the missions so they could be under the control of the government instead of the religious orders. As David Weber observes, this "new method of spiritual government" meant that "mission-held properties were to revert to the Indians; former mission Indians would come under the authority of civil officials and lose the exemption from taxes, tithes, and other fees the missions had afforded them; the *cura* would be paid by tithes from Indian parishioners rather than by the crown" (2005: 108).

29 The Spanish quote reads "Desde el año 1637 y hasta el presente son 54 los Religiosos que han muerto en las Montañas del Peru a manos de los infieles."

30 The specific location depicted in the map is what is considered the Cordillera Oriental of the Andes. Loreto surrounds it on the north, Cuzco and Junin at the south, Brazil at the east, and Huanuco, Pasco, and Junin at the south. Pampa del Sacramento constitutes the region between the Huallaga and Ucayali rivers, and is approximately 300 miles long.

31 For a discussion on the coexistence between Enlightenment ideas and religion in the case of Spain and Spanish America, see Rodríguez García (2006).

32 It is known that the *Mercurio peruano* reached cities such as La Paz, Quito, La Havana, Santa Fe de Bogota, Mexico City, Philadelphia, and countries such as Spain, Italy, Poland, Britain, France, Hungary, and Germany, attesting to the transnational and international circulation of this newspaper. For more information on this see Clément (1997: 72, 268).

12 Documentary as a space of intuition

Luis Buñuel's *Land Without Bread*

Joan Ramon Resina

Being an extension of the eye, the movie camera presupposes the body as the ultimate referent of sensory experience. Like the body, the camera is both an object and the apparatus through which a determination of space comes to pass. Both body and camera center the world. Phenomenologically, the eye is both a fleshy organ that can be rubbed, caressed, or simply seen, and the locus of sight, a condition that gives rise to two conceptions of space. In the first, bodies act upon each other, directed by an intentionality that remains outside the arena of action. This space, which I call space of ocularity, can be defined as a realm of instrumentality. It is the space where the camera manipulates object relations and is in turn manipulated, i.e., resisted by the objects with which it shares this space. Although the distances actually covered may be great, this is, in essence, a space of proximity. Here objects are grasped in their momentary emergence into an intentional field of action organized by the interplay of time and space. The location of objects within such an intentionally organized intersection is the region.

Things are different in the second kind of space, which I shall call the space of intuition. In this space, things possess all their sensorial qualities and also those aspects that are not sensory rigorously speaking, i.e., which belong to the thing without entering the field of perception. In the words of Elisabeth Ströker (1988: 85), "It is the unity of what is perceived and what is grasped along with it that first grounds the conception of the 'thing' as an identity in the fluctuation of perceptual manifolds." What is important for my purpose is that in the space of intuition, things levitate, are removed from their context of action, and leave behind their instrumental relations to other things – which regionalize and thus contextualize their scope – attaining an identity through their co-presence with a subject that provides them with continuity in their transition from one region of perception to another. A thing becomes self-identical when it transcends the context of its emergence into agency with the help of a co-present consciousness. In this consciousness's intentionality the thing retains its historical configuration – the circumstances of its spread in space as it came perceptually into being – even as it enters a future that is not yet perceived. Although the two spaces overlap in practice, analytically they are discrete moments of consciousness. One corresponds to the perception of something prior that is not yet recognized (the thing has been grasped in all its perceptual features), the other to the moment of identification. At this second stage the thing retains the perceived features in an intentional state of consciousness that mediates between the subject's past involvement in the perceptual field and the next one.

An embodied subject centers the space of intuition, but this subject's relation to space is no longer of proximity, as it was in the space of ocularity, but of remoteness. In the space of ocularity, the body is projected toward things, the *pragmata*, which emerge within a horizon of perception defined by the senses. This is a space tightly knit by the eye. The space of intuition is a space of remoteness, not just because of the distance between the subject and things, but because the former is abstracted from its immediate relation to things, withdrawn into a space beyond corporeality. The space of intuition hinges on the remoteness of the body to itself by virtue of sight. As a part of the body, the eye is an object of perception, but as the locus of sight it can never be an object. Sight's elusiveness to sensory appropriation is what makes vision the antechamber of abstraction, the turnstile, so to speak, between the bodily sphere of action and the ghostly – though not disembodied – realm of thought.

For film, especially documentary, the distinction between the space of action and the space of intuition is paramount. To be of interest, film must suggest the co-presence of the viewing subject to a constellation of *pragmata*, of things captured in their emergence within a web of relations. Sought, handled, acquired, given, stolen, or destroyed, things in film are always caught sight of at the moment of their appearance within a bound horizon of experience. Not aloof, like the spectral images of photography, filmed things exist within a context-rich environment, which renders them at once actual and contingent.

At the same time, though, film presupposes the remoteness of the spectatorial eye from the immediacy of the interchanges that define the space of action, a remoteness denoted by the camera, behind which the subject hides from view, as if removing itself to the dark chamber of vision. The camera's presence on the scene of action both removes the viewing subject from the scene and implies its attendance. Expressed in a different way, the camera's co-presence with the *pragmata*, its own status as a kind of *doing*, is the secret of cinema's trance-like power of revelation. Hence, comparison of the camera to the transcendental subject, based on its organization of data into spatiotemporal constellations, is belied by its emplacement among the things in the world. The camera's active implication permits it to function as a locative device, emplacing the spectator within the specific co-ordinates of a given representation. At the crossroads of action and intuition, the camera is the gateway through which the viewer's intentionality relates not to eidetic reductions of experience but to regional space. Unlike photography, the movie camera cannot easily disembed its objects from the system of relations in which they are found, or assign them to an ontology that abstracts them from the conditions of visual space.

This is why, typically, panoramic shots at the beginning of a movie gird the world with a horizon. Such establishing shots reproduce the kinetic feeling that accompanies the embodied perception of landscape. By tracing the limits of the world, they orient sight and bind it to the possibility of movement toward that limit. As a result, the spectator straddles the roles of sedentary voyeur lodged in the space of intuition and of the traveler who moves in space and time by slipping into the space of ocularity through identification with the eye that once stood corporeally behind the viewfinder.

Film trades in perceptual relocations. A corollary of this observation is that film shares with geography the task of producing orientation in space. Perhaps one had better say it shares this perspective with human geography, since film orients by injecting cultural and ethical lineaments into the space of ocularity. Like medieval map-makers, modern filmmakers trace the divide between civilization and primitivism, venturing into unknown territory where monsters lurk. The dimming of the lights in movie theaters suggests not so much descent into the unconscious as alternation between day and night, that is, transition from one region of the earth in which people go about their ordinary business to another, where it is other people who go about their business while the first are at rest. One could say that film formalizes the emergence of the second order observer and that the dimming of the lights is the ritual whereby the first order observer is deactivated so that the second may come into play.

Notwithstanding the idea that cinema is a projection of our unconscious desires, it should not be forgotten that the magical lantern illuminates a fragment of the world, in the sense in which medieval artists illuminated parchments and, in the baroque era, painters illuminated interiors with landscapes. In this respect fictional cinema does not differ in essence from documentary. From the point of view of visuality, it is indifferent whether cinema mimics or reproduces the *pragmata* within a given horizon. Also indifferent is whether fictionalization affects space (as in the adventure or exotic film), time (as in the historical or futuristic film), or both (as in the fantastic film, or in science fiction). What counts is that film places the eye in a non-participatory relation to the field of action, a move that centers the field, anchors its objects inside a frame of reference, and pries them loose from their contingent relations, at once redeeming them from perspectival fragmentation and dissolution in the flow of changing perceptions. But to achieve this goal, the camera must first enter the space of ocularity and become entangled with the objects lurking there.

The region that film opens up to intuition is determined by the camera's co-presence. The fact that the motion of the camera originates elsewhere, namely in an off-frame pragmatic context, allows film to modify its intended region. The camera, in other words, negotiates the conditions of vision with objects of equal ontological status and yet appears to produce those very objects as if it gave birth to the visual world. This is film's magic, a myth that is nowhere as intense as in the practice of social realism. In this sense, cinema can be likened to the cave dwellers' attempts to reproduce the hunt on the walls of their ill-lit natural palaces. By removing themselves from the actual hunting grounds, those primitive illuminators transposed the space of action to a place and into a form that they could control. Film is a venturing forth into the world through the path of sight, and documentary (film's earliest expression) is an illuminated travelogue of that exploration.

A great deal of theory has emphasized film's artificiality, its construction through editing and other techniques. Semiologically, says Michael Renov (1993: 2), there is no difference between documentary and fiction film; what distinguishes them is "the differing historical status of the referent." And yet, if no formal ground allows us to distinguish between these two sorts of film, what marks some films as documentary and others as fictional? Viewers do not need to leave the cinema or look away from their TV screens to ascertain "the status of the referent." Rather, that status appears to be a byproduct of their ability to discern the genre of the film they are watching. Documentary leads to the space of intuition, while fiction film opens onto the space of fantasy. It is familiarity with these experiences of space that discloses the historical, or better yet, the ontological status of the referent to the viewer.

Even when critics stress film's indexicality and consider every movie a "document" of the thing-in-itself (Nichols 2001: 1), that is, a sort of phenomenological schemata of an otherwise inaccessible reality, they neglect the camera's ability to break open the cocoon of idealism by calling forth not a phantasmagoria of disembodied shapes but the traces of a presence. Film, and *a fortiori* documentary, are predicated on the co-presence of a body inscribed with a visual orientation and of the viewers who, by sharing that orientation, become contemporaries in a region of experience.

What does it mean, then, that documentary determines a region and endows it with depth through the approach of the camera? In its classic age, documentary was an extension of the anthropological (and indeed colonial) voyage. For example, Flaherty returned from Alaska with a yarn about an exotic race, but the success of his tale was due to the illusion that an audience situated thousands of miles away could see Nannook through the peephole of the camera. The screen was the inner edge of a remote space, from which the spectator could look on whatever emerged within an artificial horizon. But documentary does not satisfy just by collecting objects for perception; its main interest lies in its ability to convert a space of ocularity into a space of intuition, surpassing the sensory modalities of apprehension and gathering a multiplicity of images into a unity of sense. Documentary achieves this goal by transferring a large tract of space of perception to the narrower parameters of categorical apprehension, a move epitomized by the camera's outward motion toward the world and by the countermotion that translates the world's perceptual incommensurability into images fitted to the space of the theater screen.

Documentary has certain equivalence to the task of geography inasmuch as both set out to create an epistemic representation of the world that is supported by pictures. The interlocking of film and geography can be seen to advantage in Luis Buñuel's *Land Without Bread*, a documentary that lies within a tradition of cultural reconnaissance. Its predecessors are the nineteenth century excursions of upper-class students to rural areas in search of the national spirit, which they held could be learned "intuitively" from direct observation of the landscape, peasants, their crafts, and age-old traditions. In Spain, "land-discovering" activities were promoted by the Institución Libre de Enseñanza, whose educational philosophy permeated the Residencia de Estudiantes, where Buñuel lived as a student. There is little question that Buñuel's journey to Spain's most backward region owed much to this tradition, although, as Jordana Mendelson (1996) has shown, it was also inspired by Maurice Legendre's 1927 study *Las Jurdes: Étude de géographie humaine*, a subtitle that Buñuel adopted for his film. "Étudier cet étrange pays c'est donc étudier l'Espagne elle-même," had written Legendre (Mendelson 1996: 234), and Buñuel, in turn, lifted the region to a commentary on the character of Spain, a land of extreme, dramatic contrasts at the edge of European civilization.

Las Hurdes had long been an emblem of Spain's decadence and exceptionality within Europe. Nineteenth-century critical journalist Mariano José de Larra witheringly referred to his stultified contemporaries as inhabitants of Las Batuecas, a valley in the province of Salamanca, where Buñuel's film begins. In the summer of 1914, Miguel de Unamuno wandered into Las Hurdes in the company of Maurice Legendre and Jacques Chevalier. His account of the hiking trip prefigures Buñuel's imagery, while his insistence on the inadequacy of the scientific approach to Hurdano life recalls Legendre's choice of the term "human geography." Unamuno (1922) highlights some motifs that later reappear in the documentary: the woody Batuecas valley where the ruins of a convent can be seen; the gloomy slopes and ravines in the central Hurdes; the slate-roofed stone houses with no openings except for a door; the fern-filled beds; the minuscule gardens and potato patches often destroyed by natural enemies; the tiny goats; the lack of bread in the Hurdano diet; the stunted, goiter-stricken people; the hellish streets from whose hovels occasionally the beautiful face of a small girl emerges; the contrast between the beauty and liveliness of the children and the decrepit, prematurely aged grown-ups; the public school teacher carrying on a civilizing mission against all odds, and children able, despite their misery, to read; even the macabre touch of human remains decomposing by the wayside with incongruous shreds of newspaper nearby, a surrealist tribute to Hurdano culture. And, as if to contradict Buñuel, the mention of song rising from the bottom of the ravine.

Unamuno's purpose was to show that the region's notorious backwardness was merely an acute instance of Castilian austerity, and thus an honor, rather than a dishonor, to Spain. Above all, he documents the presence of a strong cultural drive that is hemmed in by a hostile nature. The meagerness of the Hurdanos' material civilization was for him proof of their spiritual development, and their attachment to the land was evidence of the racial individualism that Unamuno placed at the heart of his cultural axiology at the time.

Buñuel, on the contrary, implied that cultural and material deprivation go hand in hand and are linked to isolation and inertia. The blame is subtly shifted to the urban centers where Unamuno locates the foci of culture and thus the primary object of administrative concern, while claiming that it is useless to try to raise the civic level of the countryside. To critique this viewpoint at a time when the Spanish government seemed incapable of implementing a badly needed land reform was, beyond simple orientation, the primary purpose of the maps shown at the beginning of the film. Appearing in succession, three maps gradually approach

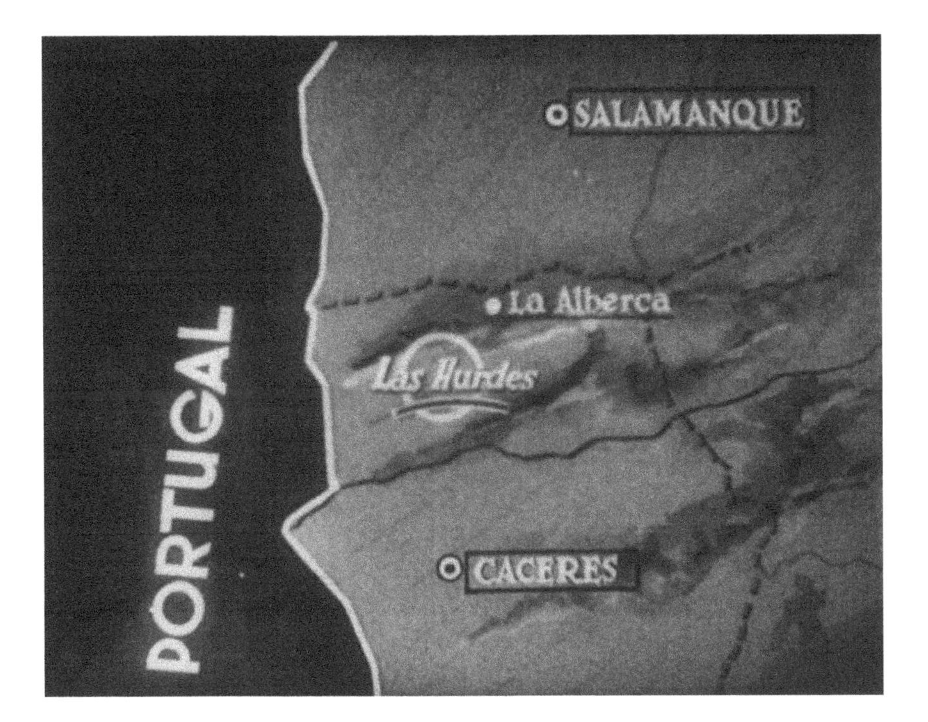

Figure 12.1 A regional map locating Las Hurdes.

the space of ocularity in a geographic regression from Europe to Spain and then to a regional close-up locating Las Hurdes between the provincial capitals of Salamanca and Cáceres, precisely the kind of centers that according to Unamuno exerted a beneficial influence on the surrounding territory (Figure 12.1).

Land Without Bread is, to paraphrase Walter Benjamin (1969), a thistle in the land of liberal thinking, which is doubtless the reason why the film was banned by the Republican government in 1934. But the point I wish to make is not that documentary is always involved in persuasion, or that it is a form of *poesis*, or that this film was inspired by surrealism, or that Buñuel looked beyond the cinematographic principles of the period, which, according to Hans Richter (1986: 42), demanded "the documented fact". Rather I wish to comment on the visual articulation of the region that gives the film its title, or rather both of its titles, as if to underscore the two kinds of space that emerge from the visual continuum.

The film is, in effect, a study in human geography, and, as soon becomes apparent, also an anthropological commentary on historical decline. The documentary opens not with a routine establishing shot but with a cartographic reference. Highlighting the immediacy of the glum to the sublime in the third, regional map, Buñuel is not merely underscoring a surrealist irony or denouncing a contrast (which might be read as a quip against Unamuno, rector of Salamanca University

Figure 12.2 Mountain range of Las Hurdes.

at that time) but suggesting the transitions implied by spatial coexistence. The panning shot, showing a lush landscape said to encircle the valley of the damned, traces a conceptual horizon, which defines but in no way limits the space about to become visible (Figure 12.2). The camera in fact does not open the spectacle in *medias res* but enters the region from a preambular space, which functions as visual threshold, and represents the last post of civilization. But what civilization?

At La Alberca, a town located nearly halfway between Salamanca and Cáceres, the filming team stops to record the persistence of ancient superstitions and callous rituals of manly prowess. The camera prepares us for what is to come by displaying the traces of a higher culture in the ruins of church towers and monasteries, from whose pinnacles the high-angle shots plummet to the ground to reveal the world of vermin (Figure 12.3). The visual plunge recalls at once the fall from paradise and the decline of a once brilliant culture, a theme over which nineteenth-century Spanish intellectuals agonized. One is reminded of their lament about Castilian decadence when Pierre Unik's voice-over informs that the Hurdanos lack not only the techniques of baking bread or ventilating a dwelling but even the essential knowledge of song. Antonio Machado's verse about hamlets of "atónitos palurdos sin danzas ni canciones" (astounded churls without dance or song) comes to mind (1991: 102). The transitional stage of La Alberca links both sides of its ocular space, enfolding them into one and the same space of intuition. Salamanca and Las Hurdes are joined in spatial adjacency; are they

Figure 12.3 Seventeenth-century religious buildings.

not also links in the same historical process? Was Spain's cultural decadence not signified by this academic redoubt of traditionalism? Was not Las Hurdes, as Legendre suggested, the country's spatial unconscious, now transposed by Buñuel into hallucinating images of human geography? This theme appears to be the meaning of the film's Dantean overtones.

A culture consisting of theology and canonical law, did it not survive in academic vestiges as useless as the magic trinkets displayed by women in La Alberca? To put it cartographically, does the map mark out spatial discontinuity or an expanded boundary for the resolution of problems posed by perception? For Tom Conley (1987: 179),

> the film implies that the voice cannot be dissociated from repression of optical splendor. The pictogrammatical element avers to be the film's unconscious; it is evident, clear, and immediately accessible. When disaster or plight is reported (in the British accent of the colonizer), oblivious to what is being said of them, children smile at the camera. They contradict the anthropological project of redemption.

Conley disengages the space of ocularity from the space of intuition, which seeks to unify the numerous visual fragments resulting from exceedingly short takes. It

Figure 12.4 Dead donkey covered by bees.

is true, in this sense, that the visual remains below the film's consciousness, which creates a surface of meaning barely detachable from colonial hermeneusis. But there is no redemption. The film traces a geography of death with the cool detachment of Dante sketching the circles of the damned.

Is not the sense of cultural deprivation, which injects meaning into the contingent images of what we experience as human catastrophe, an effect of the geography first established by the map? If the musical background to the images of wretchedness, Brahms's Fourth Symphony, is so disquieting, it is not because it represents "high culture" and "it is our culture, that of the 'educated' middle classes who watch the film" (Hopewell 1986: 19). Rather, it is disquieting because, having no naturalistic value (it does not signify "Spain"), it suggests heterogeneity that is belied by the spatial transitions between the maps.

The two most constructed sequences in the film, the donkey stung to death by bees (Figure 12.4) and the goat tumbling down to its death, are not merely instances of harsh life but illustrations of the death principle, Thanatos, that lies at the heart of every purpose caught in this stage of subsistence (Figure 12.5). At one point we see village men leave for the harvest season in Extremadura, only to return dejected a few days later. Other seasonal workers were faster on the scene, and the Hurdanos will have to wait a full year for a new chance to improve their families' diet. Yet what prevails is resigned attachment to a form of life consumed by the satisfaction of the most basic bodily requirements. Another

Figure 12.5 Falling goat.

sequence sharpens the contradiction to the utmost. In the school, barefoot children learn the rudiments of culture (Figure 12.6), the same, we are told, taught to all children in Spain. This sense of equality is all the more perturbing in that the children, who look alert and lively, anticipate the preternaturally aged men and women who represent their fate. The literacy instilled in the school (an ideal of the Spanish republic) is as pointless in the face of impending death as the injunction to respect private property, which the rudiments of culture are meant to guarantee (Figure 12.7).

The proximity of death spreads a pall of indolence over the villagers, suggesting that they are not members of a budding society but exponents of broad cultural decline. There are intimations that a catastrophe has befallen these people. Like the bee-stung donkey or the stumbling goat, the Hurdanos appear to have been thrown by a historical cataclysm to the bottom of their ravines, where even the art of agriculture is nearly forgotten in a cultural regression of thousands of years.

With the theme of degeneracy organizing the visual field, we are not surprised by the sequence of the country dwarves or the emergence of a monster-like creature from below the horizon. Commentators have identified an allusion to the Spanish tradition of painting represented by Velásquez's portraits of dwarves in the court of Philip IV. But the allusion is hardly an erudite reference or an

Figure 12.6 Barefoot children at school.

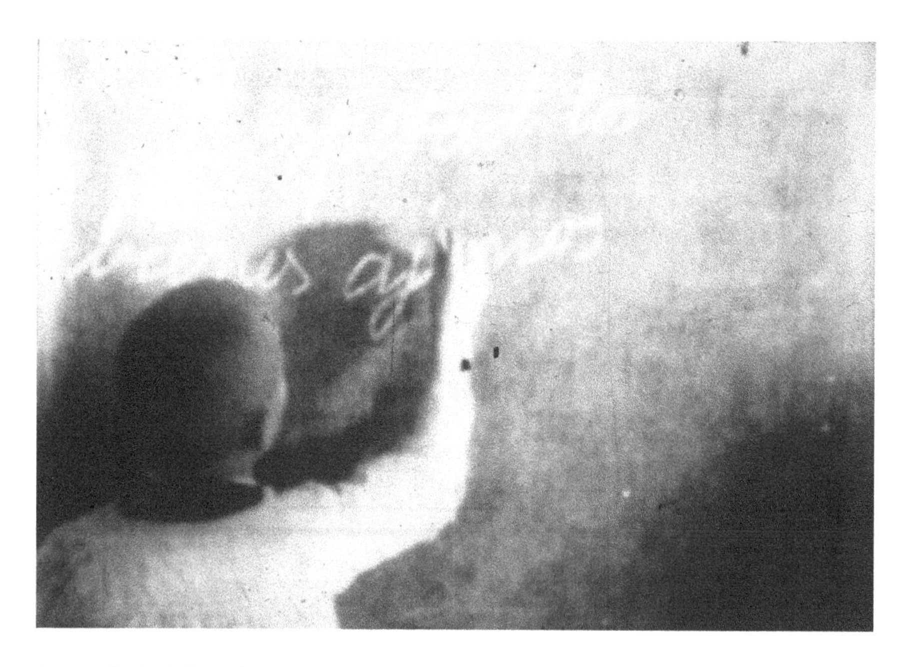

Figure 12.7 Calligraphy lesson.

expressionist gesture (Kinder 1993: 289). Buñuel seeks out dramatic opportunities, but does not resort either to classic expressionist techniques or to the clash of shots in the tradition of montage. If this scene reveals a genealogy, it is the realism represented by Velasquez, namely a focused presentation of whatever appears in the visual field. Emphasizing the surreal backdrop (Conley 1987: 178) is helpful only to the extent that surrealism relied on pictorial precision to depict images of decay. The dead donkey on the piano in *Un chien andalou* (1928) and the rotting corpses of bishops in *L'Age d'Or* (1930) stick out of the flow of images in those films, but the static, timeless image of the putrid donkey in *Las Hurdes* pervades the entire film, as if its stench saturated every sequence.

Surrealism was an assault on the mind through a jamming up of the senses. Shock results from sensory congestion. The notorious opening sequence of *Un chien andalou* relies as much on the tactile as on visual experience. Its discomfort originates in the overpowering of vision by pain vicariously felt at seeing the razor move across the eye of the actress. Similarly, *Las Hurdes* depends on visually mediated tactile experience to suggest co-presence and thus the mode of veracity. In an extraordinary compression of the visual field we are thrust into the mouth of a girl suffering from inflammation of the esophagus, so advanced, we are told, that she died in a matter of days. Although the camera does not capture her death, reporting the event with the same impotence that pervades the life of the Hurdanos stirs unease in the viewer. And the anxiety sticks to the image, as we look past the girl's bad teeth into her throat where death lurks (Figure 12.8). It is as if Buñuel wanted us to feel a pain the camera cannot visually convey. The diseased throat calls up another image of a woman suffering from goiter, so that intrusion into the girl's open mouth (in the film's space of ocularity) tantalizes with the anticipation of discovering the secret of Hurdano misery (to be resolved in the space of intuition).

Anatomical convergence of the afflicted throats of girl and woman, effected by means of rhyming shots, might be taken for a form of dialectic reminiscent of montage, but this is hardly the case. The method is not opposition and contrast so much as resonance and the dogged pursuit of an essential leitmotif. The film's overpowering tactility comes from the jagged, scraggy quality of the shots themselves. Conley (1987) counted 238 shots in 29 minutes of film. This yields a brisk rhythm, which we barely notice, however, on account of the cumulative rather than dialectical relation between the images, and also because each sequence appears complete unto itself, with the result that, as Conley puts it, "Buñuel forces the viewer to compress the effects of the long take into disgruntling rapidity" (1987: 177–8). As a result of speed, the viewer feels the flow of life as a staccato sequence of experiences arising from the haste of the camera. Without the smoothness of the long take, the film feels like ocular sandpaper.

By stressing touch, Buñuel underscores the materiality of objects that vision alone would render in two-dimensional flatness. We feel the swollen thyroids on the woman's neck, measure the size of the dwarves against ours, and stand with lowered heads in the squat, turtle-like huts on the hillside. We escort the dead baby as it is ferried over the rustic Stygian stream by the funeral cortege. And if

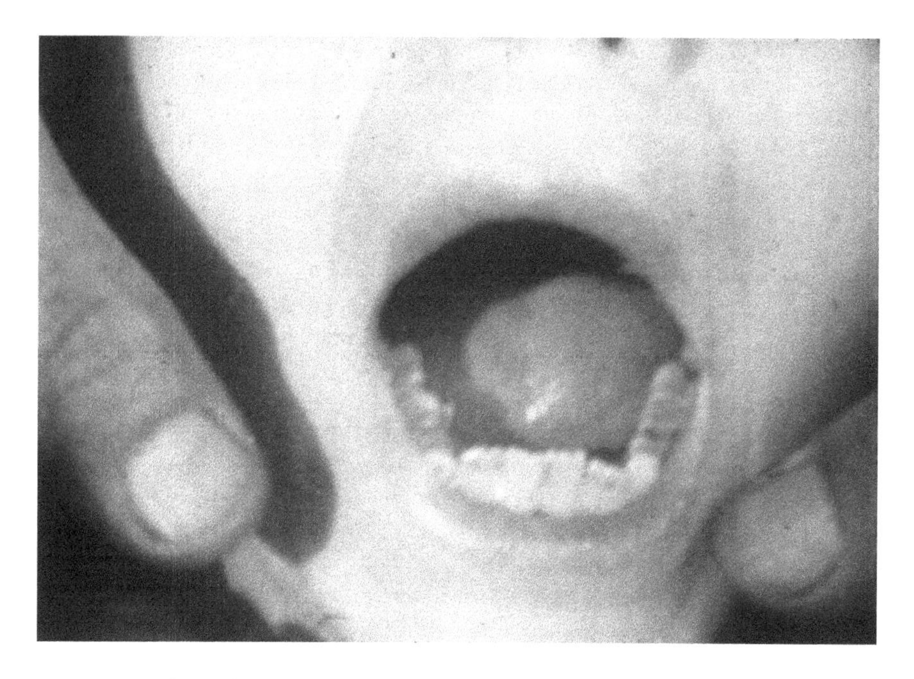

Figure 12.8 Filmmaker's hands holding girl's chin.

the donkey's putrefaction and the fall of the goat are contrived, they nevertheless merge seamlessly with the film's recurrent theme and add to its texture. Without the constant reminders of death, the film would seem almost bucolic, but it is in the presence of finitude that we become aware that this ocular space is neither arbitrary nor detachable from geographic location. Buñuel constructs the region as a bounded expanse of things that change position without altering the overall significance of the space. Even when the Hurdanos leave the region in search of work, we do not see them break through their narrow vital confines. We simply lack images of Hurdano life outside the Dantean circle of death, and cannot visually transcend its limiting finality. Of course, what we see is only what Buñuel chooses to show, but his choice depends on the delimitation of a horizon which is not only geographic but is also the sum of things visually determined by the space in which they are located.

This determining space is the space of intuition, which embraces both visual and co-given factors. In this space, the objects of vision are extricated from the relations of immediacy and alternate with what is known, presumed, or remembered. Spatiality becomes geography and geography merges with history. The filming team introduces not only a visual but also a temporal horizon, producing the shock of anachronistic contemporaneity characteristic of anthropological documentary. From the co-presence of technology and material backwardness arises a haunting impression of decadence. It does not stem from the relations of contiguity between objects in ocular space, but from the unity obtained by these images in the eyes of an external witness who is also their interpreter. The Hurdanos, in other words, are not themselves decadent; they become a striking image of decadence in the particular space of human geography.

The external witness is a cinematic function ranging from the disembodied voiceover to the implicit co-presence of the viewer, who adds his own "commentary" to the images in his intentional mode of participation. This witness processes the strictly visual data into a more complex sense formation. The contrast between the naturalistic quality of the images and the dramatic soundtrack corresponds to the sharp transitions between shots, evoking the raw, crude life with which the film – itself of poor quality – wants to acquaint us. If we ask what is it, precisely, that Buñuel's film documents, the contrast between the bucolic setting and the harsh human existence comes immediately to mind. We are in the presence of a damned paradise and a fallen race.

Who are the Hurdanos? Descendants, we are told, of bandits, Jews and heretics who fled into this remote mountains to escape the Inquisition; people trapped in a time warp. The historical background, rather than the immediate reality of cretinism in the space of ocularity, determines the artistic resonance between the scene of dwarves and paintings of similar creatures in the court of Philip IV. An allusion to decadence at the heart (or the throat) of an imperial giant afflicted with congenital deficiencies? Perspective can easily slip into causality, as we shift from space of ocularity to space of intuition. Thus, geographic adjacency, illustrated in the cartographic preamble to the film, sets off a ripple effect. From the flat horizontality of the map, the camera's descent through the threshold of La Alberca into the ravines of Las Hurdes hints at secular decadence engulfing the entire region and the country. If the point of entry to the Dantean vision of abnormal development is made coextensive with the Spain of the Hapsburgs through shots of religious ruins from the seventeenth century, the point of exit clinches the overpowering sense of arrested history through the Baroque *memento mori* proclaimed by an old woman in the empty streets at nightfall.

The camera generates ocular space by turning objects into images, but objects resist it, not yielding easily to manipulation. The presence of the filming team in Hurdano country disrupts native life, as its realism succumbs to the temptation to exacerbate its effects. Whether it be the cloud of smoke blowing from right of frame into the photogram of the stumbling goat (thus betraying the gunshot that fells the nimble animal), the solemn ferrying of an infant corpse over water, the "fortuitous" stinging to death of a beehive-laden donkey, or the extradiegetic invasion of the frame by the hand that holds the sickly girl's chin (Figure 12.9), a number of non-perceptual factors intervene that are not mere supplements to the images but wield agency in their production. In this way the phenomenological purity of the film is compromised, as is the space of intuition that gathers the multiple imprints of the objects in our senses into ontological unity. We simply cannot be sure of the identity or even self-identity of what we see. Is the girl really dying? Is the donkey's carcass the same object we saw moving briskly a few frames back? But the gesture

Figure 12.9 Sick girl amd member of filming team.

of explicit self-referentiality, by which the camera poses as actor as well as stand-in for the transcendent viewer, replicates realism's highest feat. As in Velásquez's *Meninas*, it produces the illusion of mimesis and hints at the technique by which the illusion is crafted, while embodying the eye that grasps the scene.

References

Adorno, R. (1988) *Guaman Poma: Writing and Resistance in Colonial Peru*, Austin, TX: University of Texas Press.

Agrawal, A. (2002) 'Indigenous knowledge and the politics of classification', *International Social Science Journal* 173: 287–97.

'AHR conversation: on transnational history' (2006) *American Historical Review* 3: 1441–64.

Ainsworth, W.H. (2007 [1839]) *Jack Sheppard*, ed. E. Jacobs and M. Mouro, Broadview Editions, Peterborough, ON: Broadview Press.

Albert, S. (2003) 'Critical cartography', in Furthertxt. (http://www.furthertxt.org/saulalbert.html). Revised version http://twenteenthcentury.com/saul/cartography.htm#footrev-15-0

Albertson, D. and King, C. (eds) (2008) *Without Nature: A New Condition for Theology*, New York, NY: Fordham University Press.

Aldrich, R. (2007) 'Homosexuality and the city: an historical overview', in A. Collins (ed.), *Cities of Pleasure: Sex and the Urban Socialscape*, London: Routledge, 89–108.

Ali, M. (2003) *Brick Lane: A Novel*, New York, NY: Scribner.

Allen, M. (ed.) (1997) *Ideas that Matter: The Worlds of Jane Jacobs, Owen Sound*, Toronto: Ginger Press.

Alpers, S. (1987) *El arte de describir: El arte holandés en el siglo XVII*, Madrid: Hermann Blume.

Al Sayyad, N. (2006) *Cinematic Urbanism: A History of the Modern from Reel to Real*, London: Routledge.

Anderson, B. (1983) *Imagined Communities: On the Origins and Spread of Nationalism*, London: Verso.

Anderson, J. (1996) 'The shifting stage of politics: new medieval and postmodern territorialities?', *Environment and Planning D: Society and Space* 14: 133–53.

Appadurai, A. (1988) 'Place and voice in anthropological theory', *Cultural Antropology* 3(1): 16–200.

Appadurai, A. (1996) *Modernity at Large: Cultural Dimensions of Globalization*, Minneapolis MN: University of Minnesota Press.

Arenal, E. and Schlau, S. (1989) *Untold Sisters: Hispanic Nuns in their own Works*, Albuquerque, NM: University of New Mexico Press.

Arias, S. (2007) 'Recovering imperial space in Juan Bautista Muñoz's *Historia del Nuevo Mundo*', *Revista Hispánica Moderna* 60(2): 125–41.

Arias, S. and Meléndez, M. (2002) *Mapping Colonial Spanish America: Places and Commonplaces of Identity, Culture, and Experience*, Lewisburg, NY: Bucknell University Press.

Armitage, D. (1995) 'The New World and British historical thought', in K. O. Kupperman (ed.) *America in European Consciousness 1493–1750*, Chapel Hill, NC: University of North Carolina Press.
Auden, W.H. (1940) 'As I walked out one evening', *Another Time*, London: Faber & Random House.
Bachelard, G. (1994) *The Poetics of Space*, tr. M. Jolas, Boston, MA: Beacon Press.
Baker, A.R.H. (2003) *Geography and History: Bringing the Divide*, Cambridge: Cambridge University Press.
Balibar, E. (2004) *We, the People of Europe? Reflections on Transnational Citizenship*, Princeton, NJ: Princeton University Press.
Barnes, T. (1991) *Writing Worlds: Discourse, Text and Metaphor in the Representation of Landscape*, London: Routledge.
Bartolovich, C. (1996) 'Mapping the spaces of capital', in R. Paulston (ed.), *Social Cartography: Mapping Ways of Seeing Social and Educational Change*, New York, NY: Garland Publishing.
Basso, K. (1996) 'Wisdom sits in places: notes on a Western Apache landscape', in S. Feld and S. Basso (eds), *Senses of Place*, Santa Fe, CA: School of American Research Press.
Bauman, Z. (2000) *Liquid Modernity*, Oxford: Blackwell.
Benach, N. and Albet, A. (2008) *Edward W. Soja: La perspectiva postmoderna de géografo radical*, Barcelona: Icaria.
Benjamin, Walter (1969) 'The work of art in the age of mechanical reproduction,' in H. Arendt (ed.), *Illuminations*, tr. Harry Zohn, New York, NY: Schocken.
Bergel, M. (2002) 'Lo local, lo global, lo múltiple', in *Cuadernos de la Resistencia Global*, Buenos Aires: Intergalaktika.
Berger, J. (1972) *Ways of Seeing*, London: British Broadcasting Corporation.
Berger, J. (1974) *The Look of Things*, New York, NY: Viking Press.
Berger, J. (1984) *And our Faces, my Heart, Brief as Photos*, New York: Pantheon.
Bergmann, S. (2007) 'Theology in its spatial turn: space, place, and built environments changing and challenging the images of god', *Religion Compass* 1: 1–27.
Berque, A. (1994) *Cinq propositions pour une théorie du paysage*, Paris: Champ Vallon.
Berquist, J. (2002) 'Critical spatiality and the construction of the ancient world', in D. Gunn and P. McNutt (eds), *Imagining Biblical Worlds: Studies in Spatial, Social, and Historical Constructs in Honor of James W. Flanagan*, London: Sheffield Academic Press.
Bingham, N. (1996), 'Object-ions: from technological determinism towards geographies of relations', *Environment and Planning D: Society and Space* 14: 635–57.
Bingham, N. and Thrift, N. (2000) 'Some new instructions for travelers: the geography of Bruno Latour and Michael Serres', in M. Crang and N. Thrift (eds), *Thinking Space*, London: Routledge.
Black, J. (2000) *Maps and History: Constructing Images of the Past*, New Haven, CT: Yale University Press.
Blair, S. (1998) 'Cultural geography and the place of the literary', *American Literary History* 10(3): 544–67.
Blake, E. (2002) 'Spatiality past and present: an interview with Edward Soja', *Journal of Social Archeology* 2: 139–58.
Blomley, N. (1994) *Law, Space, and the Geography of Power*, New York, NY: Guilford Press.
Blunt, A. and Rose, G. (eds) (1994) *Writing Women and Space: Colonial and Postcolonial Geographies*, New York, NY: Guilford Press.

Borges, J.L. (1981) *Historia universal de la infamia*, Barcelona: Alianza Editorial.

Botero, S. (1986) 'La geografía sagrada: la montaña de los hermanos mayores', unpublished manuscript, Bogota: Instituto Colombiano de Antropología.

Bourdieu, P. (1977) *Outline of a Theory of Practice*, Cambridge: Cambridge University Press.

Bowles, S. and Gintis, H. (1986) *Democracy and Capitalism: Property, Community, and the Contradictions of Modern Social Thought*, New York, NY: Basic Books.

Brauch, J., Lipphardt, A. and Nocke, A. (eds) (2008) *Jewish Topographies: Visions of Space, Traditions of Place*, Aldershot: Ashgate Publishing.

Brenner, N. (1998) 'Between fixity and motion: accumulation, territorial organization, and the historical geography of spatial scales', *Environment and Planning D: Society and Space* 16: 459–81.

Bruno, G. (2007) *Atlas of Emotion: Journeys in Art, Architecture, and Film*, London: Verso.

Bunce, V. (1994) 'Should transitologists be grounded', *Slavic Review* 54(1): 111–27.

Bunge, W. (1969) 'The first years of the Detroit geographical expedition: a personal report', in R. Peet (ed.), *Radical Geography: Alternative Viewpoints on Contemporary Issues*, Chicago, IL: Marroufa Press.

Bureau d'Études-Université Tangente (2002) 'Autonomous knowledge and power in a society without affects', at the Université Tangente homepage, www.utangente.free.fr. Tr. Brian Holmes. http://utangente.free.fr/anewpages/holmes.html

Bureau d'Études-Université Tangente (2003) 'Governmentality of information', at the Université Tangente homepage, www.utangente.free.fr. Tr. Brian Holmes. http://utangente.free.fr/anewpages/govinfo.html

Bureau d'Études-Université Tangente, 'Présentation: manifeste de l'Université Tangente', at the Université Tangente, http://utangente.free.fr/anewpages/pres.html

Burleigh, M. (1988) *Germany Turns Eastward: A Study of* Ostforschung *in the Third* Reich, Cambridge: Cambridge University Press.

Büttner, M. (1973) *Die Geographia generalis vor Varenius: Geographisches Weltbild und Providentialehre*, Wiesbaden: Franz Steiner Verlag GmbH.

Büttner, M. (1980) 'Survey article on the history and philosophy of religion in Germany', *Religion* 10: 86–119.

Cadavid, G. and Herrera, L.F. (1985) 'Manifestaciones culturales en el área Tayrona', *Informes Antropológicos* 1: 5–54.

Cagaptay, S. (2006) *Islam, Secularism and Nationalism in Modern Turkey: Who is a Turk?*, London: Routledge.

Cameron, D. (2007) 'Post-communist democracy: the impact of the European union', *Post-Soviet Affairs* 23(3): 185–217.

Camic, C. (ed.) (1995) *Reclaiming the Sociological Classics: The State of Scholarship*, Oxford: Blackwell.

Cañizares-Esguerra, J. (2001) *How to Write the History of the New World: Histories, Epistemologies and Identities in the Eighteenth Century Atlantic World*, Stanford, CA: Stanford University Press.

Carbia, Rómulo D. (1943/2004) *Historia de la leyenda negra hispano-americana*, Madrid: Nuevos Mundos.

Castells, M. (1977) *The Urban Question: A Marxist Approach*, Cambridge, MA: MIT Press.

Castells, M. (1996) *The Information Age*, vol. 1: *The Rise of the Network Society*, Oxford: Blackwell.

Castells, M. (1997) *The Information Age*, vol. 2: *The Power of Identity*, Oxford: Blackwell.

Castro Klarén, S. and Millones, L. (eds) (1990) *El retorno de las huacas: Estudios y documentos sobre el Taki Onqoy, siglo XVI*, Lima: Instituto de Estudios Peruanos.

Castro-Gómez, S. (2005) *La hybris del punto cero: Ciencia, raza e ilustración en la Nueva Granada (1750–1816)*, Bogota: Editorial Pontificia Javeriana.

CCAI – Centro de Coordinación de Acción Integral (2006) *Por la recuperación social del territorio*, Bogota: Presidencia de la República.

Christie, A. (1956) 'Genteel Queen of Crime: Agatha Christie puts her zest for life into murder', profile by N. Dennis, *Life* magazine (14 May).

Clarke, D. (ed.) (1997) *The Cinematic City*, London: Routledge.

Clarke, G. (1992) *Space, Time and Man: A Prehistorian's View*, Cambridge: Cambridge University Press.

Clément, J. (1997) *El Mercurio peruano, 1790–1795*, vol. 1, Madrid: Iberoamericana.

Clifford, J. (1988) *The Predicament of Culture: Twentieth-Century Ethnography, Literature and Art*, Cambridge, MA: Harvard University Press.

Clifford, J. (1997) *Routes: Travel and Translation in the Late Twentieth Century*, Cambridge, MA: Harvard University Press.

Columbus, C. (1966) *Epistola de Insulis Nuper Inventis*, tr. Frank E. Robbins, Ann Arbor, MI: March of America Facsimile Series.

Conley, T. (1987) 'Documentary surrealism: on *Land without Bread*', in Rudolf E. Kuenzli (ed.) *Dada and Surrealist Film*, New York, NY: Willis Locker & Owens.

Cosgrove, D. (1984) *Social Formation and Symbolic Landscape*, London: Croom Helm.

Cosgrove, D. (1988) 'The geometry of landscape: practical and speculative arts in sixteenth-century Venetian land territories', in D. Cosgrove and S. Daniels (eds) *The Iconography of Landscape*, Cambridge: Cambridge University Press.

Cosgrove, D. (1999) 'Introduction', in D. Cosgrove (ed.) *Mappings*, London: Reaktion Books.

Cosgrove, D. (2001) *Apollo's Eye: A Cartographic Genealogy of the Earth in the Western Imagination*, Baltimore, MD: Johns Hopkins University Press.

Cox, K. (ed.) (1997) *Spaces of Globalization: Reasserting the Power of the Local*, New York, NY: Guilford.

Crang, M. (2000) 'Public space, urban space and electronic space: would the real city please stand up?', *Urban Studies* 37: 301–17.

Crang, M., Crang, P. and May, J. (eds) (1999) *Virtual Geographies: Bodies, Space and Relations*, London: Routledge.

Cresswell, T. (2004) *Place: A Short Introduction*, Oxford: Blackwell.

Cresswell, T. (2006) *On the Move: Mobility in the Modern Western World*, New York, NY: Routledge.

Dahms, H. (2005) 'Globalization as hyper-alienation: critiques of traditional Marxism as arguments for basic income', *Social Theory as Politics in Knowledge: Current Perspectives in Social Theory* 24: 205–76.

Dahms, H. (2008) 'How social science is impossible without critical theory: the immersion of mainstream approaches in time and space', forthcoming in 'No Social Science without Critical Theory', *Current Perspectives in Social Theory* 25.

Dalla Bernardina, S. (1996) *L'Utopie de la Nature*, Paris: Imago.

Darden, K. and Grzymała-Busse, A. (2006) 'The great divide: literacy, nationalism, and the communist collapse', *World Politics* 59(1): 83–115.

Davis, Wade (2004) 'Los Kogui, guardianes del mundo', *National Geographic en español*, November 2004: 30–40.

De Angelis, M. (2003) 'Reflections on alternatives, commons and communities', *The Commoner* 6: 1–14.

De Certeau, M. (1974/1984) *The Practice of Everyday Life*, tr. S. Rendall, Berkeley, CA: University of California Press.

Deffontaines, P. (1948) *Géographie et Religions*, Paris: Gallimard.

DeGuzmán, M. (2005) *Spain's Long Shadow: The Black Legend, Off-whiteness, and Anglo-American Empire*, Minneapolis, MN: University of Minnesota Press.

De Landa, M. (1996) 'Markets, antimarkets and network economies', http://www.to.or.at/delanda/a-market.htm

De Landa, M. (2005) 'Beyond the problematic of legitimacy: military influences on civilian society', *Boundary 2* 32(1): 117–28.

Delaney, D., Ford, R. and Blomley, N. (eds) (2001) *The Legal Geographies Reader: Law, Power, Space,* Oxford: Blackwell.

Delanty, G. (2000) *Modernity and Postmodernity: Knowledge, Power, and the Self,* Thousand Oaks, CA: Sage.

Deleuze, G. (1988) *Foucault*, Minneapolis, MN: University of Minnesota Press.

Deleuze, G. (1990) 'Society of control', in *L'autre journal* 1 (May), http://www.nadir.org/nadir/archiv/netzkritik/societyofcontrol.html

Deleuze, G. and Guattari, F. (1980/1987) *A Thousand Plateaus: Capitalism and Schizophrenia* tr. B. Massumi, Minneapolis, MN: University of Minnesota Press.

Denevan, W.M. (1992) 'The pristine myth: the landscape of the Americas in 1492', *Annals of the Association of American Geographers* 82(3): 369–85.

Derrida, J. (1994) *Specters of Marx: The State of the Debt, the Work of Mourning, and the New International*, London: Routledge.

Descola, Ph. (1985) 'De l'Indien Naturalisé à l'Indien Naturaliste: sociétés amazoniennes sous le regard d'occident', in A. Cadoret (ed.) *Protection de la nature: Histoire et idéologie, De la nature à l'environnement,* Paris: L'Harmattan.

Descola, Ph. (1986) *La Nature Domestique,* Paris: MSH.

Descola, Ph. (2005) *Par-delà Nature et Culture*, Paris: Gallimard.

Deutsche, R. (1996) *Evictions: Art and Spatial Politics,* Cambridge, MA: MIT Press.

Dianteill, E. (2002) 'Deterritorialization and reterritorialization of the Orisha religion in Africa and the New World (Nigeria, Cuba and the United States)', *International Journal of Urban and Regional Research* 26(1): 121–37.

Dicken, P. (2003) *Global Shift: Reshaping the Global Economic Map in the 21st Century*, London: Guilford Press.

Dicken, P. (2007) *Global Shift: The Internationalization of Economic Activity*, 5th ed., New York, NY: Guilford Press.

Dicken, P., Kelly, P., Olds, K. and Yeung, H. (2001) 'Chains and networks, territories and scales: towards an analytical framework for the global economy', *Global Networks* 1: 89–112.

Dickens, C. (1987) *The Oxford Illustrated Dickens*, vol. 3: *Bleak House*; vol. 10: *Our Mutual Friend*, Oxford: Oxford University Press.

Dikec, M. (2001) 'Justice and the spatial imagination', *Environment and Planning A* 33: 1785–1805.

Dilley, R. (ed.) (1999) *The Problem of Context*, Oxford: Bergham Books.

Dimendberg, E. (1997) 'From Berlin to Bunker Hill: urban space, late modernity, and film noir in Fritz Lang's and Joseph Losey's M.', *Wide Angle* 19(4): 62–93, http://muse.jhu.edu/journals/wide_angle/v019/19.4dimendberg.html#fig04 (accessed 8 Oct. 2007).

DiPalma, G. (1990) *To Craft Democracies: An Essay on Democratic Transitions*. Berkeley, CA: University of California Press.

Dodge, M. and Kitchin, R. (2004) 'Flying through code/space: the real virtuality of air travel', *Environment and Planning A* 36: 195–211.

Dodge, M. and Kitchin, R. (2005a) 'Code and the transduction of space', *Annals of the Association of American Geographers* 95: 162–80.

Dodge, M. and Kitchin, R. (2005b) 'Codes of life: identification codes and the machine-readable world', *Environment and Planning D: Society and Space* 23: 851–81.

Dodge, M. and Kitchin, R. (2007) 'The automatic management of drivers and driving spaces', *Geoforum* 38: 264–75.

Dodgshon, R. (1998) *Society in Time and Space: A Geographical Perspective on Change*, Cambridge: Cambridge University Press.

Doel, M. (1999) *Poststructuralist Geographies: The Diabolical Art of Spatial Science*, Boulder, CO: Rowman & Littlefield.

Douglas, M. (1982) *Natural Symbols: Explorations in Cosmology*, New York, NY: Pantheon.

Douglas, M. (2002) *Purity and Danger: An Analysis of the Concepts of Pollution and Taboo*, London: Routledge.

Dreiser, T. (1900/1991) *Sister Carrie*, Oxford: Oxford University Press.

Duranti, A. and Goodwin, C. (eds) (1992) *Rethinking Context: Language as an Interactive Phenomenon*, Cambridge: Cambridge University Press.

Dyos, H.J. and Wolff, M. (eds) (1973) *The Victorian City: Images and Realities*, 2 vols, London: Routledge & Kegan Paul.

Edgerton, S. (1975) *The Renaissance Rediscovery of Linear Perspective*, New York, NY: Icon.

Edney, M. (1997) *Mapping an Empire: The Geographical Construction of British India*, Chicago, IL: University of Chicago Press.

Egan, P. (1821) *Life in London, or, The Day and Night Scenes of Jerry Hawthorn, Esq. And his Elegant Friend Corinthian Tom, accompanied by Bob Logic, the Oxonian, in their Rambles and Sprees through the Metropolis*, London: Sherwood, Neely, & Jones.

Egg, A.L., *Past and Present*, oil on canvas support: 635 × 762 mm frame: 801 × 925 × 85 mm painting, Tate Collection. No. 1 1858.

Eich, J. (2004) *The Other Mexican Muse: Sor María Anna Agueda de San Ignacio, 1695–1756*, New Orleans, LA: University Press of the South.

Eisenstein, E. (1979) *The Printing Press as an Agent of Change*, New York, NY: Cambridge University Press.

Elden, S. (2004) *Understanding Henri Lefebvre: Theory and the Possible*, London: Continuum.

Elden, S. and Crampton J.W. (eds) (2007) *Space, Knowledge and Power: Foucault and Geography*, London: Ashgate.

Eliade, M. (1956/1967) *Le Sacré et le Profane*, Paris: Gallimard.

Eliade, M. (1961) *The Sacred and the Profane: The Nature of Religion*, tr. Willard R. Trask, New York: Harper Torchbooks.

Elkins, D. (1995) *Beyond Sovereignty: Territory and Political Economy in the Twenty-First Century*, Toronto: University of Toronto Press.

Ellen, Roy (1993) 'Persistence and change in the relationship between anthropology and human geography', *Progress in Human Geography* 12(2): 229–62.

Escobar, A. (2001) 'Culture sits in places: reflections on globalism and subaltern strategies of localization', *Political Geography* 20: 139–74.

Escobar, A. (2004) 'Beyond the Third World: imperial globality, global coloniality, and anti-globalization social movements', *Third World Quarterly* 25(1): 207–30.

EZLN. (2005) 'Sixth Declaration of the Selva Lacandona', http://www.ezln.org/documentos/2005/sexta1.en.htm

Fabian, J. (1983) *Time and the Other: How Anthropology Makes its Object*, New York, NY: Columbia University Press.

Featherstone, D. (2003) 'Spatialities of transnational resistance to globalization: the maps of grievance of the inter-continental caravan', Transactions of the Institute of British Geographers 28: 404–21.

Featherstone, D. (2004) 'Review essay: Making trans-national spaces in politics', *Political Geography* 23(5): 625–36.

Featherstone, M. (ed.) (1990) *Global Culture: Nationalism, Globalization and Modernity*, London: Sage.

Feld, S. and Basso, S. (eds) (1996) *Senses of Place*, Santa Fe, CA: School of American Research Press.

Fickeler, P. (1962) 'Fundamental questions in the geography of religions', in P. Wagner and M. Mikesell (eds), *Readings in Cultural Geography*, Chicago, IL: University of Chicago Press.

Filler, D.M. (2004) 'Silence and the racial dimension of Megan's Law', *Iowa Law Reviews* 89: 1535–94. Available at SSRN: http://ssrn.com/abstract=648261 (accessed 2 Jan. 2007).

Fish, S. (1998) 'Democratization's requisites: the postcommunist experience', *Post-Soviet Affairs* 14(1): 212–47.

Fish, S. (1999) 'The determinants of economic reform in the post-communist world', *East European Politics and Societies* 12(2): 31–78.

Flanagan, J. (1999) 'Ancient perceptions of space/perceptions of ancient space', *Semeia*, 87: 15–43.

Flanagan, J. (2001) 'The trialectics of biblical studies', presidential address, Annual Meeting of the Eastern Great Lakes Biblical Society, Wheeling, WV.

Florida, R. (2002) *The Rise of the Creative Class*, New York, NY: Basic Books.

Foucault, M. (1972) *The Archaeology of Knowledge*, New York, NY: Pantheon.

Foucault, M. (1980) 'Questions on geography', in C. Gordon (ed.) *Power/Knowledge: Selected Interviews and Other Writings, 1972–1977*, London: Harvester Wheatsheaf.

Foucault, M. (1984) *Space, Knowledge, and Power*, in P. Rabinow (ed.) *The Foucault Reader*, New York, NY: Pantheon.

Foucault, M. (1986) 'Of other spaces', tr. J. Miskowiec, *Diacritics* 16(1): 22–7.

Frug, G. (1993) 'Decentering decentralization', *University of Chicago Law Review* 60: 253–338.

Fundación Pro Sierra Nevada (1987) 'La Fundación Pro Sierra Nevada de Santa Marta', promotional brochure, Bogota: Banco de Occidente.

Fundación Pro Sierra Nevada (1988a) 'Presentación', *Diagnóstico de la Sierra Nevada de Santa Marta*, Santa Marta: Gobernación del Magdalena, Corporación Regional del Cesar, Corporación Regional de la Guajira.

Fundación Pro Sierra Nevada (1988b) 'Contribución a una historia oral de la colonización de la SNSM, por Alfredo Molano', *Diagnóstico de la Sierra Nevada de Santa Marta* (see 1988a).

Fundación Pro Sierra Nevada (1988c) 'La cuestión indígena y la situación de las fronteras de los resguardos por Silvia Botero y Camilo Arbeláez', in *Diagnóstico de la Sierra Nevada de Santa Marta* (see 1988a).

Gamble, C. (1987) 'Archeology, geography and time', *Progress in Human Geography* 11: 227–46.

García Lorca, F. *Poeta en Nueva York*, http://bivir.uacj.mx/LibrosElectronicosLibres/Autores/FedericoGarciaLorca/Poeta%20en%20Nueva%20York.pdf. (accessed 28 Nov. 2007).

Gensler, M. (1999) 'The doctrine of place in a commentary on the "physics" attributed to Antonius Andeae', *Early Science and Medicine* 4: 329–58.

Gerbi, A. (1973) *The Dispute of the New World: The History of a Polemic, 1750–1900*, tr. J. Moyle, Pittsburgh, PA: University of Pittsburgh Press.

Gereffi, G. (1996) 'Global commodity chains: new forms of coordination and control among nations and firms in international industries', *Competition and Change* 1: 427–39.

Gereffi, G. and Korzeniewicz, M. (eds) (1994) *Commodity Chains and Global Capitalism*, Westport, CT: Praeger.

Gergen, K. (1991) *The Saturated Self: Dilemmas of Identity in Contemporary Life*, New York, NY: Basic Books.

Gibson, C. (1971) *Black Legend: Anti-Spanish Attitudes in the Old World and the New*, New York, NY: Random House.

Gibson, C. (2003) 'New economic geographies', in B. Garner (ed.), *Geography's New Frontiers, The Geographical Society of New South Wales Conference Papers no. 17*, Sydney: University of New South Wales.

Gibson-Graham, J.K. (2001) 'An ethics of the local', http://www.communityeconomies.org/papers/rethink/rethinkp1.pdf

Gibson-Graham, J.K. (2002) 'A diverse economy: rethinking economy and economic representation', http://www.communityeconomies.org/papers/rethink/rethink7diverse.pdf

Gibson-Graham, J.K. (2003) 'Feminizing the economy: metaphors, strategies, politics', *Gender, Place and Culture* 10(2): 145–57.

Gibson-Graham, J.K. (2005) *A Post-Capitalist Politics*, Minneapolis, MN: University of Minnesota Press.

Giddens, A. (1984) *The Constitution of Society: Outline of the Theory of Structuration*, Berkeley, CA: University of California Press.

Gil, R. (1994) Interview held in the village of Kemakúmeke, Guachaca River, between 27 March and 3 April.

Gil, R. (2007) Entrevista realizada con Juana Londoño en Santa Marta entre Junio y Julio. Interview by Juana Londoño in Santa Marta in June and July.

Gilbert, P.K. (2004) *Mapping the Victorian Social Body*, Albany, NY: SUNY Press.

Gilpin, R. (1987) *The Political Economy of International Relations*, Princeton, NJ: Princeton University Press.

Gómez-Peña, G. (1993) *Warriors for Gringostroika*, St Paul, MN: Greywolf Press.

Gottdiener, M. (1985) *The Social Production of Urban Space*, Austin, TX: University of Texas Press.

Gottdiener, M. (2002) 'Urban analysis as merchandising: the "LA School" and the understanding of metropolitan development', in J. Eade and C. Mele (eds) *Understanding the City*, Oxford: Blackwell.

Gouldner, A. (1985) *Against Fragmentation: The Origins of Marxism and the Sociology of Intellectuals*, New York, NY: Oxford University Press.

Graeber, D. (2002a) 'Marcel Mauss: Give it away', Interactivist Info Exchange, http://slash.autonomedia.org/print.pl?sid=02/10/11/1246214

Graeber, D. (2002b) 'The new anarchists', *New Left Review* 13: 61–73; http://www.newleftreview.net/NLR24704.shtml.

Graeber, D. (2004) *Fragments of an Anarchist Anthropology*, Chicago: Prickly Paradigm Press.

Graham, S. (1999) 'Global grids of glass: on global cities, telecommunications, and planetary urban networks', *Urban Studies* 36: 929–49.

Graham, S. and Aurigi, A. (1997) 'Virtual cities, social polarization, and the crisis in urban public space', *Journal of Urban Technology* 4: 19–52.

Graham, S. and Marvin, S. (2001) *Splintering Urbanism: Networked Infrastructures, Technological Mobilities and the Urban Condition*, London: Routledge.

Gregory, D. (1994) *Geographical Imaginations*, Oxford: Blackwell.

Gubrium, J. and Holstein, J. (2000) *Institutional Selves: Troubled Identities in a Postmodern World*, New York, NY: Oxford University Press.

Gupta, A. and Fergusson, J. (eds) (1992a) 'Theme Issue: Space, Identity and the Politics of Difference', *Cultural Anthropology* 7(1).

Gupta, A. and Fergusson, J. (1992b) 'Beyond "Culture": Space, Identity and the Politics of Difference', *Cultural Anthropology* 7(1): 6–23.

Gupta, A. and Fergusson, J. (1997) *Anthropological Locations: Boundaries and Grounds of a Field Science*, Berkeley, CA: University of California Press.

Habermas, J. (1992) *The Structural Transformation of the Public Sphere*, tr. T. Burger with F. Lawrence, Cambridge: Polity Press.

Haggett, P. (1965) *Locational Analysis in Human Geography*, London: Edward Arnold.

Hand, C.M. and Judkins, B. (1999) 'Disciplinary schisms: subspecialty "drift" and the fragmentation of sociology', *The American Sociologist* 30(1): 18–36.

Hanson, S.E. (1995) 'The Leninist Legacy and Institutional Change', *Comparative Political Studies* 28(2): 306–14.

Haraway, D. (1991) *Simians, Cyborgs, and Women: The Re-invention of Nature*, London: Free Association Books.

Harley, J.B. (1989) 'Deconstructing the map', *Cartographica* 26: 1–20.

Harley, J.B. (1992) 'Rereading the Maps of the Columbian Encounter', *Annals of the Association of American Geographers* 82(3): 522–42.

Harrison, R. (1992) *Forêts: Essai sur l'imaginaire occidental,* Paris: Flammarion.

Hartog, F. (1991) *Le miroir d'Hérodote: Essai sur la représentation de l'autre*, Paris: Gallimard.

Hartwick, E. (1998) 'Geographies of consumption: a commodity-chain approach', *Environment and Planning D: Society and Space* 16: 423–37.

Harvey, D. (1973) *Social Justice and the City*, Baltimore, MD: Johns Hopkins University Press.

Harvey, D. (1982) *The Limits to Capital*, Chicago, IL: University of Chicago Press.

Harvey, D. (1984) 'On the history and present condition of geogarphy: an historical materialist manifesto', *The Professional Geogapher* 36: 1–11.

Harvey, D. (1985) *The Urbanization of Capital*, Oxford: Blackwell.

Harvey, D. (1989) *The Condition of Postmodernity*, Oxford: Blackwell.

Harvey, D. (1990) 'Between space and time: reflections on the geographical imagination', *Annals of the Association of American Geographers* 80: 418–34.

Harvey, D. (1995) 'Geographical knowledge in the eye of power: reflections on Derek Gregory's *Geographical Imaginations*', *Annals of the Association of American Geographers* 85: 160–4.

Harvey, D. (2006a) 'Neoliberalism as creative destruction', *Geografiska Annaler* 88B: 145–58.

Harvey, D. (2006b) *Spaces of Global Capitalism: Towards a Theory of Uneven Geographical Development*, London: Verso.

Hervieu-Léger, D. (2002) 'Space and religion: new approaches to religious spatiality in modernity', *International Journal of Urban and Regional Research* 26: 99–105.

Hill, R. (2000) *Sceptres and Sciences in the Spains: Four Humanities and the New Philosophy (ca. 1680–1740)*, Liverpool: Liverpool University Press.

Hill, R. (2005) *Hierarchy, Commerce, and Fraud in Bourbon Spanish America: A Postal Inspector's Exposé*, Nashville, TN: Vanderbilt University Press.

Hillis, K. (1999) 'Toward the light "within": optical technologies, spatial metaphors, and changing subjectivities', in M. Crang, P. Crang and J. May (eds) *Virtual Geographies: Bodies, Space and Relations*, London: Routledge.

Hirsch, E. (1995) 'Introduction: landscape: between place and space', in E. Hirsch and M. O'Hanlon (eds), *The Anthropology of Landscape, Perspectives on Place and Space*, Oxford: Clarendon Press.

Hirsch, E. and O'Hanlon, M. (eds) (1995) *The Anthropology of Landscape: Perspectives on Place and Space*, Oxford: Clarendon Press.

Hodder, I. (2006) *The Leopard's Tale: Revealing the Mysteries of Catalhoyuk*, London: Thames & Hudson.

Hodder, I. and Orton, C. (1979) *Spatial Analysis in Archeology*, Cambridge: Cambridge University Press.

Holloway, J. (2002) *Change the World without Taking Power*, London: Pluto Press.

Holmes, B. (2002) 'Mapping excess, seeking uses: Bureau d'Études and multiplicity', at the Université Tangente homepage, www.utangente.free.fr. http://utangente.free.fr/anewpages/cartesholmes2.html

Holmes, B. (2003) 'Maps for the outside: Bureau d'Études, or the revenge of the concepts', Interactivist Info Exchange, http://info.interactivist.net/print.pl?sid=03/10/10/0141258

Holmes, B. (2004a) 'Imaginary Maps, Global Solidarities', at the Piet Zwart Institute, http://pzwart.wdka.hro.nl/mdr/pubsfolder/bhimaginary/

Holmes, B. (2004b) 'Flowmaps, the imaginaries of global integration', at the Piet Zwart Institute, https://pzwart.wdka.hro.nl/mdr/pubsfolder/bhflowmaps/

hooks, b. (1984) *Feminist Theory: From Margin to Center*, Boston, MA: South End.

Hopewell, J. (1986) *Out of the Past: Spanish Cinema After Franco*, London: British Film Institute.

Howard, M. (2003) *The Weakness of Civil Society in Post-Communist Europe*, Cambridge: Cambridge University Press.

Humphreys, L. (1970) *Tearoom Trade: Impersonal Sex in Public Places*, London: Duckworth.

Humphreys, R.A. (1969) 'William Robertson and his *History of America*', in R.A. Humphreys (ed.), *Tradition and Revolt in Latin America and Other Essays*, New York, NY: Columbia University Press.

Ibsen, K. (1999) *Women's Spiritual Autobiography in Colonial Spanish America*, Gainesville, FL: University of Florida Press.

Irigaray, L. (1985) *This Sex Which is Not One*, tr. C. Porter, Ithaca, NY: Cornell University Press.

Isard, W. (1972) *Location and Space-Economy: A General Theory Relating to Industrial Location, Market Areas, Land Use, and Urban Structure*, Cambridge, MA: MIT Press.

Isin, E. (2002) *Being Political: Genealogies of Citizenship*, Minneapolis, MN: University of Minnesota Press.

Isin, E. (2006) 'Space', in B. Turner (ed.), *The Cambridge Dictionary of Sociology*, Cambridge: Cambridge University Press.

Jacob, C. (2006) *The Sovereign Map: Theoretical Approaches in Cartography throughout History*, tr. T. Conley, ed. E. Dahl, Chicago, IL: University of Chicago Press.

Jacobs, J. (1969) *The Economy of Cities*, New York, NY: Random House.

Jameson, F. (1991/2003) *Postmodernism or, the Cultural logic of Late Capitalism*, Durham, NC: Duke University Press.

Janos, A. (1986) *Politics and Paradigms*, Stanford, CA: Stanford University Press.
Jay, M. (1993) *Downcast Eyes: The Denigration of Vision in Twentieth-Century French Thought*, Berkeley, CA: University of California Press.
Jenks, C. (1995) *Visual Culture,* London: Routledge.
Johnson, C. (1982) *MITI and the Japanese Miracle: The Growth of Industrial Policy 1925–1975*, Stanford, CA: Stanford University Press.
Johnson, P. (2002) *The Renaissance: A Short History*, New York: Modern Library.
Jones, S.S. (2001) *Durkheim Reconsidered*, Oxford: Blackwell.
Jowitt, K. (1992) *New World Disorder: The Leninist Extinction*, Berkeley, CA: University of California Press.
Juan, J. and Ulloa, A. (1983) *Noticias secretas de América*, Bogota: Biblioteca Banco Popular.
Juderías, J. (1914) *La leyenda negra: Estudios sobre el concepto de España en el extranjero*, Madrid: Tip. de la 'Rev. de Arch., Bibl. y Museos'.
Kahn, M. (1996) 'Your place and mine: sharing emotional landscapes in Wamira, Papua New Guinea', in S. Feld and S. Basso (eds), *Senses of Place*, Santa Fe, CA: School of American Research Press.
Kalberg, S. (1994) *Max Weber's Comparative Historical Sociology*, Chicago, IL: University of Chicago Press.
Kanarinka (2005) 'How to make the invisible stay invisible: three cases studies in micropolitical engineering', presented at the Workshop on Radical Empiricism: The Sinews of the Present: Genealogies of Biopolitics, http://www.radicalempiricism.org/biotextes/textes/kanarinka.pdf
Kaplan, M. (1999) 'Who's afraid of John Saul? Urban culture and the politics of desire in late-Victorian England', *GLQ: A Journal of Lesbian and Gay Studies* 5(3): 267–314, http://muse.jhu.edu/journals/journal_of_lesbian_and_gay_studies/v005/5.3kaplan.html (accessed 3 Nov. 2007).
Katsiaficas, G. (1987) *The Imagination of the New Left: A Global Analysis of 1968*, Boston, MA: South End Press.
Katsiaficas, G. (1997) *The Subversion of Politics*, Atlantic Highlands, NJ: Humanities Press.
Katsiaficas, G. (2004) *Confronting Capitalism*, New York, NY: Soft Skull Press.
Kern, S. (1983) *The Culture of Time and Space 1880–1918*, Cambridge, MA: Harvard University Press.
Kincaid, J.R. (1998) *Erotic Innocence: The Culture of Child Molesting*, Durham, NC: Duke University Press.
Kinder, M. (1993) *Blood Cinema: The Reconstruction of National Identity in Spain*, Berkeley, CA: University of California Press.
Kirby, K. (1996) *Indifferent Boundaries: Spatial Concepts of Human Subjectivity*, New York, NY: Guilford.
Kirschner, R. (1984) 'The vocation of holiness in late Antiquity', *Vigiliae Christianae* 38: 105–24.
Kitchin, R. (1998) *Cyberspace: The World in the Wires*, New York, NY: John Wiley.
Kitschelt, H. (2003) 'Accounting for postcommunist regime diversity: what counts as a good cause', in G. Ekiert and S. Hanson (eds), *Capitalism and Democracy in Central and Eastern Europe*, Cambridge: Cambridge University Press.
Knott, K. (2005) *The Location of Religion: A Spatial Analysis*, London: Equinox.
Kohn, M. (2003) *Radical Space: Building the House of the People*, Ithaca, NY: Cornell University Press.

Kohn, R. (2006) 'In and out of place: physical space and social location in the Bible', in J. Wood, J. Harvey and M. Leuchter (eds), *From Babel to Babylon: Essays on Biblical History and Literature in Honour of Brian Peckham*, London: Continuum.
Kong, L. (1990) 'Geography of religion: trends and prospects', *Progress in Human Geography* 14: 355–71.
Kong, L. (2001) 'Mapping "new" geographies of religion: politics and poetics in modernity', *Progress in Human Geography* 25: 211–33.
Kong, L. (2004) 'Religious landscapes', in J.S. Duncan, R.H. Schein, and Nuala C. Johnson (eds), *A Companion to Cultural Geography*, Oxford: Blackwell, pp. 365–81.
Kopstein, J.S. and Reilly, D. (2000) 'Geographic diffusion and the transformation of the post-communist world', *World Politics* 53(1): 1–37.
Kopstein, J. and Reilly, D. (2003) 'Post-communist spaces: a political geography approach to explaining postcommunist outcomes', in G. Ekiert and S. Hanson (eds), *Capitalism and Democracy in Central and Eastern Europe*, Cambridge: Cambridge University Press.
Koven, S. (2004) *Slumming: Sexual and Social Politics in Victorian London*, Princeton, NJ: Princeton University Press.
Kristeva, J. (1986) 'Women's time', in *The Kristeva Reader*, (ed.) T. Moi, Oxford: Blackwell, 188–213.
Kuda (2004) *Bitomatik: Art Practice in the Time of Information/Media Domination*, Novi Sad: Futura Publikacije.
Lang, F. (dir.) (1931) *M*, T. von Harbou (script) and F. Lang (script), Berlin.
Langebaeck, C. (1985) 'Tres formas de acceso a productos en territorios de los cacicazgos sujetos al Cocuy siglo XVI', *Boletín del Museo del Oro* 18: 29–45.
Langebaeck, C. (2005) 'De los Alpes a las selvas y montañas: El legado de Gerardo Reichel-Dolmatoff', *Antípoda: Revista de Antropología y Arqueología* 1: 139–71.
Lankina, T. and Getachew, L. (2006) 'A geographic incremental theory of democratization: territory, aid, and democracy in postcommunist regions', *World Politics* 58(4): 536–82.
Larsen, J., Urry, J. and Axhausen, K. (2006) *Mobilities, Networks, Geographies*, London: Ashgate.
Lash, S. and Urry, J. (1987) *The End of Organized Capitalism*, Madison, WI: University of Wisconsin Press.
Latham, A. (2002) 'Retheorizing the scale of globalization: topologies, actor-networks, and cosmopolitanism', in A. Herod and M. Wright (eds), *Geographies of Power: Placing Scale*, Oxford: Blackwell.
Latour, B. (1993) *We have Never been Modern*, London: Harvester Wheatsheaf.
Latour, B. (2004) 'A prologue in form of a dialogue between a student and his (somewhat) Socratic professor', http://www.ensmp.fr/~latour/articles/090.html
Law, J. (1994) *Organizing Modernity: Social Ordering and Social Theory*, Oxford: Blackwell.
Law, J. and Hassard, J. (eds) (1999) *Actor Network Theory and After*, Oxford: Blackwell.
Lefebvre, H. (1962) 'La significación de la comuna', archived at Espai Marx, http://www.moviments.net/espaimarx/docs/c9892a989183de32e976c6f04e700201.pdf
Lefebvre, H. (1974/1991/1998) *The Production of Space*, Oxford: Blackwell.
Lefebvre, H. (1996) *Writings on Cities*, tr. E. Kofman and E. Lebas, Oxford: Blackwell.
Legendre, M. (1927) *Las Jurdes: Étude de géographie humaine*, Bordeaux: Feret & Fils, Éditeurs.
Lemert, C. (2007) *Thinking the Unthinkable: The Riddles of Classical Social Theories*, Boulder, CO: Paradigm.

León, L. (2004) *La Llorona's Children: Religion, Life, and Death in the U.S.–Mexican Borderlands*, Berkeley, CA: University of California Press.
Levine, G. (1986) 'On the geography of religion', *Transactions of the Institute of British Geographers*, NS 11: 428–40.
Levitsky, S. and Way, L. (forthcoming) *Competitive Authoritarianism: International Linkage, Organizational Power and the Fate of Hybrid Rule*, Cambridge: Cambridge University Press.
Levitt, P. (2004) 'Redefining the boundaries of belonging: the institutional character of transnational religious life', *Sociology of Religion* 65: 1–18.
Levy, J. (1999) *Le tournant geographique*, Paris: Belin.
Liggett, H. and Perry, D. (eds) (1995) *Spatial Practices. Critical Explorations in Social/Spatial Theory*, Thousand Oaks, CA: Sage.
Linz, J. and Stepan, A. (1996) *Problems of Democratic Transition and Consolidation*, Baltimore, MD: Johns Hopkins University Press.
Lipschutz, R. (1992) 'Reconstructing world politics: the emergence of global civil society', *Millennium Journal of International Studies* 21(3): 398–420.
Lipset, S. (1963) *Political Man*, New York, NY: Doubleday.
Livingstone, D. (1984) 'Natural theology and neo-lamarckism: the changing context of nineteenth-century geography in the United States and Great Britain', *Annals of the Association of American Geographers* 74: 9–28.
Livingstone, D. (1992) *The Geographical Tradition: Episodes in the History of a Contested Enterprise*, Oxford: Blackwell.
Livingstone, D. (1994) 'Science and religion: foreword to the historical geography of an encounter', *Journal of Historical Geography* 20: 367–83.
Livingstone, D. (2003) *Putting Science in its Place. Geographies of Scientific Knowledge*, Chicago, IL: University of Chicago Press.
Loochkartt, S. and Ávila, C. (2004) *Memoria, territorio y cultura: Agua y tiempo, naturaleza y norma en dos áreas protegidas SIRAP-CAR*, Bogota: CAR.
López Baralt, M. (1988) *Ícono y conquista: Guaman Poma de Ayala*, Madrid: Hiperión.
Low, S. (2000) *On the Plaza: The Politics of Public Space and Culture*, Austin, TX: University of Texas Press.
Low, S. and Lawrence-Zúñiga, D. (eds) (2003) *The Anthropology of Space and Place*, Oxford: Blackwell.
Lucas, R. (1988) 'On the mechanics of economic development', *Journal of Monetary Economics* 22: 3–42.
Luke, T. and Ó Tuathail, G. (1998) 'Global flowmations, local fundamentalisms, and fast geopolitics: America in an accelerating world order', in A. Herod, G. Ó Tuathail and S. Roberts (eds), *An Unruly World? Globalization, Governance and Geography*, London: Routledge.
Macek, S. (2006) *Urban Nightmares: The Media, The Right, and the Moral Panic over the City*, Minneapolis, MN: University of Minnesota Press.
Machado, A. (1991) *Campos de Castilla* ed. Geoffrey Ribbans, Madrid: Cátedra.
Mackinder, H. (1904) 'The geographical pivot of history', *Geographical Journal* 23: 421–44.
Malinowski, B. (1961) *Argonauts of the Western Pacific*, New York: Dutton & Co.
Malkki, L. (1992) 'National Geographic: the rooting of peoples and the territorialization of national identity among scholars and refugees', *Cultural Anthropology* 7(1): 24–44.
Maltby, W.S. (1971) *The Black Legend in England*, Durham, NC: Duke University Press.
Mangieri, T., McCourt, M., Ruiz-Junco, N. and West, J. (2004) 'Rethinking politics, scholarship, and economics: *disClosure* interviews David F. Ruccio', *disClosure* 13: 39–64.

Marcos, Subcomandante (1997) 'The seven loose pieces of the global jigsaw puzzle', http://flag.blackened.net/revolt/mexico/ezln/1997/jigsaw.html (originally published in *Le Monde Diplomatique* (July 1997).
Marcos, Subcomandante (1999) 'The fourth world war', tr. Irlandesa, *In Motion Magazine* (Nov. 2001), http://www.inmotionmagazine.com/auto/fourth.html
Marcus, G. (1995) 'Ethnography in/of the world system: the emergence of multi-sited ethnography', *Annual Review of Anthropology* 24: 95–117.
Martin, D. and Miller, B. (2003) 'Spaces of contentious politics', *Mobilization: An International Journal*, special issue on Space and Contentious Politics 8(2): 143–56.
Martson, S. (2003) 'Mobilizing geography: locating space in social movement theory', *Mobilization: An International Journal*, special issue on Space and Contentious Politics 8(2): 227–31.
Mary, A. (2002) 'Pilgrimage to Imeko (Nigeria): an African church in the time of the "global village"', *International Journal of Urban and Regional Research* 26: 106–20.
Mason, J.A. (1936) *Archaeology of Santa Marta, Colombia: The Tayrona Culture*, Chicago, IL: Field Museum of Natural History.
Massey, D. (1991) 'Flexible sexism', *Environment and Planning D: Society and Space* 9: 31–57.
Massey, D. (1993) 'Power-geometry and a progressive sense of place', in J. Bird (ed.), *Mapping the Futures: Local Cultures, Global Change*, London: Routledge.
Massey, D. (1994) *Space, Place, and Gender*, Minneapolis, MN: University of Minnesota Press.
Massey, D. (1999) 'Imagining globalization: power-geometries of time-space', in A. Brah, M. Hickman and M. Mac an Ghail (eds), *Global Futures: Migration, Environment and Globalization*, Basingstoke: Macmillan.
Massey, D. (1999) *Power-geometries and the Politics of Space-Time, Hettner-Lecture 1998*, Heidelberg: Department of Geography, University of Heidelberg.
Massey, D. (2005) *For Space*, London: Sage.
Massumi, B. (2002) *Parables for the Virtual: Movement, Affect and Sensation*, Durham, NC: Duke University Press.
Mayr, J. (ed.) (1984a) *La Sierra Nevada de Santa Marta*, Bogota: Mayr y Cabal Editores.
Mayr, J. (1984b) 'Introducción', in J. Mayr (ed.), *La Sierra Nevada de Santa Marta*, Bogota: Mayr y Cabal Editores.
McGrane, B. (1983) *Beyond Anthropology: Society and the Other*, New York, NY: Columbia University Press.
McKnight, K. (1997) *The Mystic of Tunja: The Writings of Madre Castillo 1671–1742*, Amherst, MA: University of Massachusetts Press.
McKnight, K. (1999) 'Blasphemy as resistance: an Afro-Mexican slave woman before the Mexican Inquisition', in M. Giles (ed.) *Women in the Inquisition. Spain and the New World*, Baltimore, MD: Johns Hopkins University Press.
McKnight, K. (2003) ' "En su tierra lo aprendió": an African curandero's defense before the Cartagena inquisition', *Colonial Latin American Review* 12(1): 63–84.
McKnight, K. (2004) 'Confronted rituals: Spanish colonial and Angolan "maroon" executions in Cartagena de Indias (1634)', *Journal of Colonialism and Colonial History* 5(3): 1–23.
Meléndez, M. (2006) '*Patria, Criollos*, and Blacks: imagining the nation in the *Mercurio peruano*, 1791–1795', *Colonial Latin American Review* 15(2): 207–27.
Mellaart, J. (1964) 'A Neolithic city in Turkey', *Scientific American* 210(4): 94–104.
Melucci, A. (1989) *Nomads of the Present*, (ed.) J. Keane and P. Mier, Philadelphia: Temple University Press.

Melucci, A. (1996) *Challenging Codes: Collective Action in the Information Age*, Cambridge: Cambridge University Press.
Mendelson, J. (1996) 'Contested territory: the politics of geography in Luis Buñuel's *Las Hurdes: tierra sin pan*', *Locus Amoenus* 2: 229–42.
Midnight Notes Collective (2004) 'The new enclosures: Planetary class struggle', in D. Solnit (ed.), *Globalize Liberation*, San Francisco, CA: City Lights Books.
Mignolo, W. (1995) *The Darker Side of the Renaissance: Literacy, Territoriality, and Colonization*, Ann Arbor, MI: University of Michigan Press.
Mignolo, W. (2000) *Local Histories/Global Designs: Coloniality, Subaltern Knowledges, and Border Thinking*, Princeton, NJ: Princeton University Press.
Mignolo, W. (2005) *The Idea of Latin America*, Oxford: Blackwell.
Mignolo, W. and Boone, E. (eds) (1994) *Writing without Words: Alternative Literacies in Mesoamerica and the Andes*, Durham, NJ: Duke University Press.
Miller, B. (2000) *Geography and Social Movements*, Minneapolis, MN: University of Minnesota Press.
Mitchell, T. (1986) *Iconology: Image, Text, Ideology,* Chicago, IL: Chicago University Press.
Mogel, L. and Bhagat, A. (eds) (2007) *An Atlas of Radical Cartography*, Los Angeles, CA: Journal of Aesthetics and Protest Press.
Moretti, F. (1998) *Atlas of the European Novel*, London: Verso.
Moretti, F. (2005) *Graphs, Maps, Trees: Abstract Models for a Literary History*, London: Verso.
Mumford, L. (1934) *Technics and Civilization*, New York, NY: Harcourt Brace.
Muñoz, J.B. (1793/1990) *Historia del Nuevo-Mundo*, Madrid: Universitat de València.
Muñoz, J.B. (1797) *The History of the New World*, London: Printed for G.G. and J. Robinson.
Murdoch, J. (1997) 'Towards a geography of heterogeneous associations', *Progress in Human Geography* 21: 321–37.
Murdoch, J. (1998) 'The spaces of actor-network theory', *Geoforum* 29: 357–74.
Murdoch, J. (2006) *Post-Structuralist Geography: A Guide to Relational Space*, London: Sage.
Murra, J. (1975) *Formaciones económicas y políticas del mundo Andino*, Lima: Instituto de Estudios Peruanos.
Myers, K. (2003) *Neither Saints nor Sinners: Writing the Lives of Women in Spanish America*, Oxford: Oxford University Press.
Nabhan-Warren, K. (2005) *The Virgin of El Barrio: Marian Apparitions, Catholic Evangelizing, and Mexican American Activism*, New York, NY: NYU Press.
Naipaul, V.S. (1967) *The Mimic Men*, New York, NY: Macmillan.
Nichols, B. (2001) *Introduction to Documentary*, Bloomington, IN: Indiana University Press.
Nisbet, R. (1980) *History of the Idea of Progress,* New York, NY: Basic Books.
Nochpin, L. (1978), 'Lost and *found*: once more the fallen woman', *Art Bulletin*, 60(1) (Mar.): 139–53.
Nord, D.E. (1995) *Walking the Victorian Streets. Women, Representation, and the City*, Ithaca, NJ: Cornell University Press.
Notes From Nowhere (eds) (2003) *We are Everywhere: The Irresistible Rise of Global Anticapitalism*, London: Verso.
Oates, J.C. (1981) *Literature and the Urban Experience: Essays on the City and Literature*, ed. M.C. Jaye and A.C. Watts, New Brunswick, NJ: Rutgers University Press.

O'Donnell, G. and Schmitter P. (1986) *Transitions from Authoritarian Rule: Tentative Conclusions about Uncertain Democracies*, Baltimore, MD: Johns Hopkins University Press.

O'Gorman, E. (1958/1984) *La invención de América*, Mexico: Medio Siglo.

OGT – Organización Gonawindúa Tairona (1997) 'Palabras de mama', unpublished manuscript, Santa Marta: Proyecto Gonawindua – Ricerca e Coperazione.

Ohmae, K. (1990) *The Borderless World: Power and Strategy in the Interlinked Economy*, New York, NY: HarperCollins.

O'Loughlin, J., Flint, C. and Anselin, L. (1994) 'The geography of the Nazi vote: context, confession and class in the Reichstag election of 1930', *Annals, Association of American Geographers* 84: 351–80.

Orsi, R. (1985) *The Madonna of 115th Street: Faith and Community in Italian Harlem, 1880–1950*, New Haven, CT: Yale University Press.

Ortellado, P. (2002) 'Aproximaciones al "movimiento anti-globalización"', *Cuadernos de la Resistencia Global*, Buenos Aires: Intergalaktika.

Osborne, T. and Rose, N. (1998) *Governing Cities: Liberalism, Neoliberalism and Advanced Liberalism*, Urban Studies Programme, Working Paper, 19, Toronto: Urban Studies Programme, Division of Social Science, York University.

Ó Tuathail, G. (1996) *Critical Geopolitics*, Minneapolis, MN: University of Minnesota Press.

Paasi, A. (2004) 'Place and region: looking through the prism of scale', *Progress in Human Geography* 28: 536–46.

Padrón, R. (2004) *The Spacious Word: Cartography, Literature, and Empire in Early Modern Spain*, Chicago, IL: University of Chicago Press.

Park, C. (1994) *Sacred Worlds: An Introduction to Geography and Religion*, London: Routledge.

Paulston, R. and Liebman, M. (1994) 'The promise of critical social cartography', at the Interamerican Association for Cooperation and Development, http://www.iacd.oas.org/La%20Educa%20119/pauls.htm

Perera, V. (1998) 'Voices of Ancient Wisdom rise to save the planet from pollution', *Los Angeles Times* (19 April): M2.

Pérez de Lama, J. (2004) 'Antagonistic flows/geographies of the multitude' (Flujos antagonistas/geografías de la multitude), http://www.hackitectura.net/osfav_lados/txts/geografias.html

Phillips, B. (2001) *Beyond the Tower of Babel in Sociology: Reconstructing the Scientific Method*, New York, NY: Aldine de Gruyter.

Phillips, B. and Johnston, L. (2007) *The Invisible Crisis of Contemporary Society: Reconstructing Sociology's Fundamental Assumptions*, Boulder, CO: Paradigm.

Phillipson, N. (1997) in S.J. Brown (ed.), *William Robertson and the Expansion of Empire*, Cambridge: Cambridge University Press.

Pickles, J. (2004) *A History of Spaces: Cartographic Reason, Mapping, and the Geo-Coded World*, London: Routledge.

Pickles, J. (2006) 'On the social lives of maps and the politics of diagrams: a story of power, seduction, and disappearance. Response to review essays by J. Painter and M. Doel on *A History of Spaces: Cartographic Reason, Mapping, and the Geo-coded World*', *Area* 37: 355–64.

Pile, S. (1997) 'Opposition, political identities and spaces of resistance', in S. Pile and M. Keith (eds), *Geographies of Resistance*, London: Routledge.

Pile, S. and Keith, M. (1997) *Geographies of Resistance*, London: Routledge.

'Politics, Culture and Justice: Women and the Politics of Place', http://www.sidint.org/Publications/Leaflets/Politics_Culture_Justice.pdf
Pop-Eleches, G. (2007) 'Historical legacies and post-communist regime change', *Journal of Politics* 69(4): 908–26.
Postone, M. (1993) *Time, Labor, and Social Domination: A Reinterpretation of Marx's Critical Theory*, Cambridge: Cambridge University Press.
Postone, M. (2007) 'Theorizing the contemporary world: David Harvey, Giovanni Arrighi, Robert Brenner', in R. Albritton, B. Jessop, and R. Westra (eds), *Political Economy and Global Capitalism: The 21st Century, Present and Future*, London: Anthem Press.
Pratt, M.L. (1992) *Imperial Eyes: Travel Writing and Transculturation*, London: Routledge.
Prindle, D. (2006) *The Paradox of Democratic Capitalism. Politics and Economics in American Thought*, Baltimore, MD: Johns Hopkins University Press.
Proffitt, S. (1997) 'Jane Jacobs: still challenging the way we think about cities', *Los Angeles Times* (12 Oct.).
Prokosch, M. and Raymond, L. (eds) (2002) *The Global Activist's Manual: Local Ways to Change the World*, New York, NY: Thunder's Mouth Press.
Przeworski, A. (1991) *Democracy and the Market: Political and Economic Reforms in Eastern Europe and Latin America,* Cambridge: Cambridge University Press.
Rabasa, J. (1993) *Inventing America: Spanish Historiography and the Formation of Eurocentrism*, Norman, OK: University of Oklahoma Press.
Rabasa, J. (2000) *Writing Violence on the Northern Frontier: The Historiography of Sixteenth Century New Mexico and Florida and the Legacy of Conquest*, Durham, NC: Duke University Press.
Rabinow, P. (1977) *Reflections on Fieldwork in Morocco*, Berkeley, CA: University of California Press.
Raynal, G. (1776) *A Philosophical and Political History of the Settlements and Trade of the Europeans in the East and West Indies*, tr. J. Justamond, London: T. Cadell.
Reagan, R. 'City upon a hill' (http://www.presidentreagan.info/speeches/city_upon_a_hill.cfm, accessed 17 Sept. 2006).
Rechy, J. (1963) *City of Night*, New York, NY: Grove Press.
Reichel, G. (1947) 'Aspectos económicos entre los indios de la Sierra Nevada', *Boletín de Arqueología* 2(5–6): 573–80.
Reichel, G. (1985) *Los Kogui: Una tribu de la Sierra Nevada de Santa Marta*, vol. 1, Bogota: Procultura.
Renov, M. (1993) 'The truth about non-fiction', in Michael Renov (ed.), *Theorizing Documentary,* London: Routledge.
Richter, H. (1986) *The Struggle for the Film*, tr. Ben Brewster, New York, NY: St Martin's Press.
Rival, L. (1998) 'Domestication as an historical and symbolic process: wild gardens and cultivated forests in the Ecuadorian Amazon', in W. Balée (ed.), *Advances in Historical Ecology*, New York, NY: Columbia University Press.
Robertson, R. (1992) *Globalization: Social Theory and Global Culture,* Newbury Park, CA: Sage Publications.
Robertson, W. (1777) *History of America*, London: W. Strahan, T. Cadell & J. Balfour.
Robertson, W. (1851) *History of America*, (ed.) D. Steward, vol. 5, London: Longman, Brown, Green & Longmans.
Rodman, M. (2003) 'Empowering place: multilocalty and multivocality', in S. Low and D. Lawrence-Zúñiga (eds), *The Anthropology of Space and Place*, Oxford: Blackwell.

Rodríguez García, M. (2006) *Criollismo y patria en la Lima ilustrada (1732–1795)*, Madrid: Miño y Dávila Editores.

Roger, A. (1997) *Court traité du paysage*, Paris: Gallimard.

Romanyshyn, R. (1993) 'The despotic eye and its shadow: media image in the age of literacy', in D. Levin (ed.), *Modernity and the Hegemony of Vision*, Berkeley, CA: University of California Press.

Rorty, R. (1981) *Philosophy and the Mirror of Nature*, Princeton, NJ: Princeton University Press.

Rosenzweig, R. (1998) 'Wizards, bureaucrats, warriors, and hackers: writing the history of the internet', *American Historical Review* 103: 1530–52.

Rossetti, D.G. (begun 1853 or 1859) *Found*, oil on canvas, 36 × 31½ inches. Delaware Art Museum, Wilmington, DE.

Rossetti, D.G. (1913 [1887]) 'Jenny', in *The Poetical Works*, 2 vols, ed. W.M. Rossetti, Boston, MA: Little, Brown, 1, 110–27.

Routledge, P. (2003) 'Convergence space: process geographies of grassroots globalization networks', *Transactions of the Institute of British Geographers* 28: 333–49.

Roy, W. and Ahmed, P. (2001) 'Space', in W. Roy (ed.), *Making Societies: The Historical Construction of our World*, Thousand Oaks, CA: Pine Forge Press.

Ruggie, J. (1993) 'Territoriality and beyond: problematizing modernity in international relations', *International Organization* 47: 139–74.

Said, E.W. (1978) *Orientalism*, New York, NY: Vintage.

Saint-Blancat, C. (2002) 'Islam in diaspora: between reterritorialization and extraterritoriality', *International Journal of Urban and Regional Research* 26: 138–51.

Santiago, R. and Olkon, S. (2005) 'Molester ban poses risks, some say', *Miami Beach Herald* (19 Sept.)

Sassen, S. (2000) 'Spatialities and temporalities of the global: elements for a theorization', *Public Culture* 12: 215–32.

Sayer, D. (1991) *Capitalism and Modernity: An Excursus on Marx and Weber*, London: Routledge.

Schiller, D. (1999) *Digital Capitalism: Networking the Global Market System*, Cambridge, MA: MIT Press.

Schimmel, A. (1991) 'Sacred geography in Islam', in Jamie Scott and Paul Simpson-Housley (eds), *Sacred Places and Profane Spaces: Essays in the Geographics of Judaism, Christianity, and Islam*, New York, NY: Greenwood Press.

Schmitter, P.C. and Karl, T.L. (1994) 'The conceptual travels of transitologists and consolidologists: how far to the east should they attempt to go', *Slavic Review* 53(1): 173–85.

Schumpeter, J. (1942) *Capitalism, Socialism, and Democracy*, New York, NY: Harper.

Scott, J.C. (1998) *Seeing like a State: How Certain Schemes to Improve the Human Condition have Failed*, New Haven, CT: Yale University Press.

Serje, M. (1987a) 'Arquitectura para ciudad perdida: un modelo de asentamiento para la Sierra Nevada de Santa Marta', unpublished monograph, Bogota: Universidad de los Andes, Facultad de Arquitectura.

Serje, M. (1987b) 'Arquitectura y urbanismo en la cultura Tairona', *Boletín del Museo del Oro* 19: 87–96.

Serje, M. (2005) *El Revés de la Nación: Territorios salvajes, fronteras y tierras de nadie*, Bogota: Ediciones Universidad de los Andes.

Serres, M. (1997) *The Troubador of Knowledge*, Ann Arbor, MI: University of Michigan Press.

Serres, M. and Latour, B. (1995) *Conversations on Science, Culture and Time*, Ann Arbor, MI: University of Michigan Press.

Sex and the City, official website, http://www.hbo.com/city/ (accessed 30 Sept 2007).
Shands, K. (1999) *Embracing Space: Spatial Metaphors in Feminist Discourse*, Westport, CT: Greenwood Press.
Sheppard, E. (2002) 'The spaces and times of globalization: place, scale, networks, and positionality', *Economic Geography* 78: 307–30.
Sher, R.B. (2006) *The Enlightenment and the Book: Scottish Authors and their Publishers in Eighteenth-Century Britain, Ireland and America*, Chicago, IL: Chicago University Press.
Silber, I. (1995) 'Space, fields, boundaries: the rise of spatial metaphors in contemporary sociological theory', *Social Research* 62: 323–55.
Sinclair, U. (1981) *The Jungle*, with an introduction by M. Dickstein, New York, NY, and Toronto: Bantam.
SIP – Secretaría de Integración Popular (1990) *Plan Regional de Rehabilitación del Distrito Sierra Nevada de Santa Marta*, Bogota: Presidencia de la República.
Sipe, M. (2004) 'Romancing the city: Arthur Symons and the spatial politics of aesthetics in 1890s London', in P.K. Gilbert (ed.), *Mapping the Victorian Social Body*, Albany, NY: SUNY Press 69–85.
Slater, C. (2003) *Entangled Edens: Visions of the Amazon*, Berkeley, CA: University of California Press.
Smith, J. (1624) *The Generall History of Virginia, New-England, and the Summer Isles*, London: Printed by I.D. and I.H. for Michael Sparkes.
Smith, N. (1991) *Uneven Development: Nature, Capital and the Production of Space*, Oxford: Blackwell.
Smith, N. (2003) *American Empire: Roosevelt's Geographer and the Prelude to Globalization*, Berkeley, CA: University of California Press.
Smitten, J. (1985) 'Impartiality in Robertson's *History of America*', *Eighteenth-Century Studies* 19: 56–77.
Smitten, J. (1990) 'Moderantism and history: William Robertson's unfinished history of British America' in R. Sher and J. Smitten (eds), *Scotland and America in the Age of the Enlightenment*, Edinburgh: Edinburgh University Press.
Sobchak, V. (1997) *Screening Space: The American Science Fiction Film*, New Brunswick, NJ: Rutgers University Press.
Soja, E. (1968) *The Geography of Modernization in Kenya*, Syracuse, NY: Syracuse University Press.
Soja, E. (1971) *The Political Organization of Space*, Commission on College Geography, Resource Paper 8, Washington, DC: Association of American Geographers.
Soja, E. (1980) 'The socio-spatial dialectic', *Annals of the Association of American Geographers* 70: 207–25.
Soja, E. (1989) *Postmodern Geographies: The Reassertion of Space in Critical Social Theory*, London: Verso.
Soja, E. (1993) 'Postmodern geographies and the critique of historicism', in J. Jones, W. Natter and T. Schatzki (eds), *Postmodern Contentions*, New York, NY: Guilford.
Soja, E. (1996) *Thirdspace: Journeys to Los Angeles and Other Real-and-Imagined Places*, Oxford: Blackwell.
Soja, E. (2000) *Postmetropolis: Critical Studies of Cities and Regions*, Oxford: Blackwell.
Soja, E. (2003) 'Writing the city spatially', *City* 7: 269–80.
Soja, E. (2007) 'Postmetropolitan psychasthenia: a spatioanalysis', in BAVO (eds), *Urban Politics Now: Re-Imagining Democracy in the Neoliberal City*, Rotterdam: NAi Publishers.

Soja, E. and Kanai, J. (2008) 'The urbanization of the world', in R. Burdett and D. Sudjic (eds), *The Endless City*, London: Phaedon.

Solnit, D. (ed.) (2004) *Globalize Liberation*, San Francisco, CA: City Lights Books.

Sopher, D. (1967) *Geography of Religions*, Englewood Cliffs, NJ: Prentice-Hall.

Sorokin, P. (1964) *Sociocultural Causality, Space, Time: A Study of Referential Principles of Sociology and Social Science*, New York: Russell & Russell.

Soto Arango, D. and Puig-Samper, M.A. (eds) (1995) *La Ilustración en la América colonial*, Madrid: Consejo de Investigaciones Científicas.

Soto Arango, D. and Puig-Samper, M.A. (eds) (2003) *Recepción y difusión de textos ilustrados*, Madrid: Colección Actas Tavara, S.L.

Stallybrass, P. and White, A. (1986) *The Politics and Poetics of Transgression*, Ithaca, NY: Cornell University Press.

Steinmetz, G. (2005) 'Scientific authority and transition to post-fordism: the plausibility of positivism in U.S. sociology since 1945', in G. Steinmetz (ed.), *The Politics of Method in the Human Sciences. Positivism and its Epistemological Others*, Durham, NC: Duke University Press.

Stolley, K. (1996) 'The eighteenth century: narrative forms, scholarship, and learning', in R. González-Echevarría and E. Pupo-Walker (eds), *The Cambridge History of Latin American Literature*, vol. II, Cambridge: Cambridge University Press.

Strabo (1932) *The Geography of Strabo*, tr. Horace Leonard Jones, vol. 7, New York, NY: G. P. Putnam's Sons.

Ströker, E. (1988) *Investigations in Philosophy of Space*, tr. Algis Mickunas, Athens, OH: Ohio University Press.

Sui, D. (2000) 'Visuality, aurality, and shifting metaphors of geographical thought in the late twentieth century', *Annals of the Association of American Geographers* 90: 322–43.

Swyngedow, E. (1989) 'The heart of the place: the resurrection of locality in an age of hyperspace', *Geografiska Annaler* 71: 31–42.

Swyngedouw, E. (1997) 'Neither global nor local: "glocalization" and the politics of scale', in K. Cox (ed.), *Spaces of Globalization: Reasserting the Power of the Local*, New York, NY: Guilford.

Tally, R., Jr. (1996) 'Jameson's project of cognitive mapping: a critical engagement', in R. Paulston (ed.), *Social Cartography: Mapping Ways of Seeing Social and Educational Change*, New York, NY: Garland Publishing.

Taylor, P. (1999) *Modernities: A Geohistorical Interpretation*, Minneapolis, MN: University of Minnesota Press.

Taylor, V. (1989) 'Social movement continuity: the women's movement in abeyance', *American Sociological Review* 54: 761–75.

'Teach yourself institutions', http://www.metamute.com/look/article.tpl?IdLanguage=1&IdPublication=1&NrIssue=28&NrSection=10&NrArticle=1404

'The Rise of Network Universities', http://www.campusactivism.org/uploads/the_rise_of_network_universities.pdf

Thomas, K. (1996) *Man and the Natural World: Changing Attitudes in England 1500–1800*, Oxford: Oxford University Press (originally published 1983).

Thomas, N. (1996) *Out of Time: History and Evolution in Anthropological Discourse*, Ann Arbor, MI: University of Michigan Press.

Thompson, L. (1991) 'Mapping an apocalyptic world', in Jamie Scott and Paul Simpson-Housley (eds), *Sacred Places and Profane Spaces: Essays in the Geographics of Judaism, Christianity, and Islam*, New York: Greenwood Press.

Thrift, N. (n.d.) 'Space', unpublished mimeo. http://www2.unine.ch/documentmanager/files/lsh/geographie/EcoleDoctorale/space.pdf

Thrift, N. and Leyshon, A. (1994) 'A phantom state? The de-traditionalization of money, the international financial system and international financial centres', *Political Geography* 13: 299–327.

Tilly, C. (1998) 'Spaces of contention', Contentious Politics Working Paper Series http://www.ciaonet.org/wps/tic10/.

Tilly, C. (2003) 'Contention over space and place', *Mobilization: An International Journal*, special issue on Space and Contentious Politics 8: 221–5.

Tracy, J. (ed.) (1991) *The Political Economy of Merchant Empires*, Cambridge: Cambridge University Press.

Trouillot, M.-R. (2003) *Global Transformations. Anthropology and the Modern World*, New York: Palgrave-Macmillan.

Turner, B. (1999) *Classical Sociology*, Thousand Oaks, CA: Sage.

Turner, V. (1969) *The Ritual Process: Structure and Anti-Structure,* Chicago, IL: Aldine.

Tweed, T. (2006) *Crossings and Dwellings: A Theory of Religion*, Cambridge, MA: Harvard University Press.

Ulloa, A. (2004) *La construcción del nativo ecológico,* Bogota: ICANH–Colciencias.

Unamuno, Miguel de (1922) 'Las Hurdes', *Andanzas y visiones españolas*, Madrid: Renacimiento.

Université Tangente (2003) 'Governing by networks', http://utangente.free.fr/anewpages/archcap.html

Uribe, C.A. (1988) 'De la Sierra Nevada de Santa Marta, sus ecosistemas, indígenas y antropólogos', *Revista de Antropología y Arqueología* 4(1): 7–35.

Uribe, C.A. (2006) 'Y me citarán por muchos años más: el modelo interpretativo de Garardo Reiche-Dolmatoff y la antropología de la Sierra Nevada de Santa Marta', in A. Abello (ed.), *El Caribe en la Nación Colombiana*, Bogota: Museo Nacional de Colombia-Observatorio del Caribe Colombiano.

Urry, J. (2000) *Sociology Beyond Societies: Mobilities for the Twenty-first Century*, London and New York, NY: Routledge.

Urry, J. (2001) 'The sociology of space and place', in J. Blau (ed.), *The Blackwell Companion to Sociology*, Oxford: Blackwell.

Urry, J. (2003) *Global Complexity*, Cambridge: Polity.

Vachudova, M. (2005) *Europe Undivided: Democracy, Leverage, and Integration After Communism*, Oxford: Oxford University Press.

Van der Hammen, M.C. (1992) *El manejo del mundo: naturaleza y sociedad entre los Yukuna de la Amazonía colombiana*, Bogota: Tropenbos y Tercer Mundo Editores.

Van der Hammen, M.C. and Rodríguez, C. (2000) 'Restauración ecológica permanente: Lecciones del manejo del bosque amazónico por comunidades indígenas del medio y bajo Caquetá', in E. Ponce de León (ed.), *Memorias del Seminario de Restauración Ecológica y Reforestación*, Bogota: FAAE–Fescol–GTZ.

Van Gennep, A. (1960) *The Rites of Passage*, tr. M. Visedom and G. Caffe, Chicago, IL: University of Chicago Press.

Velazco, S. (2003) *Visiones de Anáhuac: Reconstrucciones historiográficas y etnicidades emergentes en el México colonial: Fernando de Alva Ixtlilxóchitl, Diego Muñoz Camargo y Hernando Alvarado Tezozómoc*, Guadalajara: Universidad de Guadalajara.

Verdesio, G. (2001) *Forgotten Conquests: Rereading New World History from the Margins*, Philadelphia, PA: Temple University Press.

Wagstaff, J. (ed.) (1987) *Landscape and Culture: Geographical and Archeological Perspectives*, Oxford: Blackwell.

Walker, C. (2003) 'The upper classes and their upper stories: architecture and the aftermath of the Lima earthquake of 1746', *Hispanic American Historical Review* 83(1): 53–82.

Walkowitz, J. (1992) *City of Dreadful Delight: Narratives of Sexual Danger in Late-Victorian England*, Chicago, IL: University of Chicago Press.

Warf, B. (1995) 'Telecommunications and the changing geographies of knowledge transmission in the late 20th century', *Urban Studies* 32: 361–378.

Weber, D. (2005) *Bárbaros: Spaniards and their Savages in the Ages of the Enlightenment*, New Haven, CT: Yale University Press.

Weber, M. ([1905] 2002) *The Protestant Ethic and the 'Spirit' of Capitalism, and Other Writings*, tr. P. Baehr and G. Wells, New York, NY: Penguin Books.

Wey Gómez, N. (forthcoming) *The Machine of the World: Place, Colonialism, and Columbus's Invention of the American Tropics*, Cambridge, MA: MIT Press.

Whitaker, A. (1961) *Latin America and the Enlightenment*, Ithaca, NY: Cornell University Press.

White, H. (1978) 'The historical text as literary artifact', *Tropics of Discourse: Essays in Cultural Criticism*, Baltimore, MD: Johns Hopkins University Press.

White, H. (1987) *Content of the Form: Narrative Discourse and Historical Representation*, Baltimore, MD: Johns Hopkins University Press.

Whitman, W. (1900) 'City of orgies', *Leaves of Grass*, Philadelphia, PA: David McKay.

Whitman, W. (1900) 'Once I pass'd through a populous city', *Leaves of Grass*, Philadelphia, PA: David McKay.

Winthrop, J. (1630) 'A model of christian charity', http://religiousfreedom.lib.virginia.edu/sacred/charity.html, (accessed 12 Oct. 2007).

Withers, C. (2007) *Placing the Enlightenment: Thinking Geographically about the Age of Reason*, Chicago, IL: University of Chicago Press.

Wolford, W. (2003) 'Families, fields, and fighting for land: the spatial dynamics of contention in rural Brazil', *Mobilization: An International Journal*, special issue on Space and Contentious Politics 8: 201–15.

Wood, D. (2005) 'PPGIS: Public participation? Geographic? Information? Systems?', Keynote Address at the URISA Conference on PPGIS, Cleveland, State University, Cleveland, OH, 31 July 2005.

Wood, E. (1995) *Democracy against Capitalism: Renewing Historical Materialism*, Cambridge: Cambridge University Press.

World Bank (2004) 'Documento preliminar revisado de la política sobre pueblos indígenas' (OP/BP 4.10), online, available from: http://lnweb18.worldbank.org/ESSD/sdvext.nsf/63ByDocName/DocumentoPreliminarRevisadodelaPolíticasobrePueblosIndígenas DocumentoPreliminarRevisadoOP410/$FILE/Revised Draft OP 4.10 12-01-04 SPANISH.pdf

World Social Forum (2005) 'Closing ceremony', www.forumsocialmundial.org.br/dinamic.php?pagina encerra2005_ing

World Social Forum (2005) *World Charter of the Rights to the City*, http://www.choike.org/nuevo_eng/informes/2243.html

Worthington, S. (2005) 'The Bermuda triangle in reverse: mapping contemporary capitalism', at the Riga Centre for New Media Culture, http://rixc.lv/reader/txt/txt.php?id=192&l=en

Yeung, H. (2003) ‘Practicing new economic geographies: a methodological examination’, *Annals of the Association of American Geographers* 93: 442–62.

Young, I. (1990) *Justice and the Politics of Difference*, Princeton, NJ: Princeton University Press.

Young, I. (2000) ‘Democratic regionalism’, in *Inclusion and Democracy*, New York, NY: Oxford University Press.

Zamora, M. (1993) *Reading Columbus*, Berkeley, CA: University of California Press.

Zhao, S. (1993) ‘Realms, subfields, and perspectives: on the differentiation and fragmentation of sociology’, *The American Sociologist* 24(3–4): 5–14.

Index

www.ingramcontent.com/pod-product-compliance
Ingram Content Group UK Ltd.
Pitfield, Milton Keynes, MK11 3LW, UK
UKHW021452080625
459435UK00012B/466

* 9 7 8 0 4 1 5 7 6 2 2 1 2 *